AF738469

Routledge Revivals

To The Central African Lakes and Back

To The Central African Lakes and Back

The Narrative of The Royal Geographical Society's East Central African Expedition 1878-80

Joseph Thomson

Second Edition

Volume I

Routledge
Taylor & Francis Group
Routledge Revivals

First published in 1968 by Frank Cass & Co. Ltd.

This edition first published in 2018 by Routledge
2 Park Square, Milton Park, Abingdon, Oxon, OX14 4RN
and by Routledge
52 Vanderbilt Avenue, New York, NY 10017

Routledge is an imprint of the Taylor & Francis Group, an informa business

© 1968 by Taylor & Francis

All rights reserved. No part of this book may be reprinted or reproduced or utilised in any form or by any electronic, mechanical, or other means, now known or hereafter invented, including photocopying and recording, or in any information storage or retrieval system, without permission in writing from the publishers.

Publisher's Note
The publisher has gone to great lengths to ensure the quality of this reprint but points out that some imperfections in the original copies may be apparent.

Disclaimer
The publisher has made every effort to trace copyright holders and welcomes correspondence from those they have been unable to contact.

A Library of Congress record exists under ISBN:

ISBN 13: 978-0-367-15060-0 (hbk)
ISBN 13: 978-0-429-05475-4 (ebk)

CASS LIBRARY OF AFRICAN STUDIES

TRAVELS AND NARRATIVES

No. 46

Editorial Adviser: JOHN RALPH WILLIS
Centre of West African Studies, University of Birmingham

TO THE
CENTRAL AFRICAN LAKES
AND BACK

THE NARRATIVE OF
THE ROYAL GEOGRAPHICAL SOCIETY'S
EAST CENTRAL AFRICAN EXPEDITION
1878–80

BY

JOSEPH THOMSON

SECOND EDITION
WITH A NEW INTRODUCTION BY
ROBERT I. ROTBERG

IN TWO VOLUMES

VOLUME I

FRANK CASS & CO. LTD.
1968

Published by
FRANK CASS AND COMPANY LIMITED
67 Great Russell Street, London W.C.1.

First edition	1881
Second edition	1968

Printed in Great Britain by
Thomas Nelson (Printers) Ltd., London and Edinburgh

INTRODUCTION
TO
THE SECOND EDITION

BY

ROBERT I. ROTBERG

THIS, the record of Joseph Thomson's first journey into the interior of Africa, established his credentials as an explorer of the first rank. Although he made no startling discoveries and his experiences lack the dramatic impact of those retailed in *Through Masai Land*, his more widely known and subsequent narrative, *To The Central African Lakes* added measurably to Europe's knowledge of a part of eastern Africa which hitherto had only been understood in imperfect outline. Thomson became the first European to reach the northernmost extremity of Lake Nyasa, to traverse the intervening plateaux to Lake Tanganyika, and to espy Lake Rukwa. He also pioneered a southern route from Dar es Salaam to the lakes and verified that the Lukuga River was indeed the true outlet of Lake Tanganyika. In mid-1880, a mere fourteen months after he and his party had left Zanzibar, Thomson returned there to virtually unanimous praise. He had successfully covered more than

3000 miles without once firing a shot in anger. A boy when he first set foot in Africa, he had emerged a man and a hero.

Thomson's accomplishments were appropriately recognized. Sir John Kirk, the British Consul-General at Zanzibar, wrote privately that the expedition had been "a grand success. . . . [It] shows what can be done with care—and reflects the highest credit on Thomson." Lord Aberdare, then the President of the Royal Geographical Society, voiced general sentiments: "None of the expeditions in which this Society have borne a part. . . had been marked by greater courage, by greater skill, or by greater success than this."[1] For anyone, his were impressive achievements, but for a previously inexperienced youth of 22 they were doubly so. Thomson had grown up in and around Thornhill in Dumfriesshire, where his father worked a local quarry. Young Thomson early became interested in geology and briefly studied the subject at the University of Edinburgh under Professor (later Sir) Archibald Geikie. In the summer of 1878, after leaving the University, he chanced to read in a local newspaper that the Royal Geographical Society was about to send an expedition into the African interior. Although the notice did not ask for recruits, Thomson applied to join, and the Society decided that Keith Johnston, the leader of the expedition, could well use a second-in-command who was also reasonably well qualified as a geologist. Johnston, twelve years older than Thomson, was a very experienced cartographer who had served as the

[1] Kirk to Henry Walter Bates (Assistant Secretary of the Royal Geographical Society), 20 July 1880, R.G.S. archives; Aberdare quoted in *The Proceedings of the Royal Geographical Society*, II (1880), 772.

assistant map curator of the Society and had already visited Paraguay on its behalf.[1]

From the beginning Johnston and Thomson got on poorly. Johnston thought Thomson uncouth and rough and Thomson could hardly warm to someone who was rarely more than civil. (The published narrative that follows alludes only once, and that tangentially, to this unpleasantness.[2]) Nevertheless, they travelled harmoniously to Zanzibar in late 1878, briefly visited Usambara early in the next year, and, with Kirk's assistance, prepared very carefully (Johnston was meticulous) for their pioneering journey to Lake Nyasa.

Within six weeks of their departure from Dar es Salaam, Johnston was dead of dysentery. Responsibility for the continued success of their mission devolved upon Thomson. He was young and untried, but grasped the opportunity to make his name famous and never once thought of turning back. Moreover, he knew that he could trust himself to the expert guidance of James Chuma, his leading African assistant and a sometime follower of David Livingstone. Together, as the first volume of this account makes abundantly clear, they trekked through the country of the Khutu people, avoided conflict with Mahenge warriors, crossed the Great Ruaha River in a collapsible boat, proceeded up the rich Kilombero Valley, traversed the

[1] For more detail on Thomson and Johnston and the preparations for the expedition, see Robert I. Rotberg, *Quest Without End: Joseph Thomson and the Opening-Up of Africa*, forthcoming.

[2] Kirk took the credit for preventing Thomson from making his disagreement with Johnston a public scandal. "We shall all be spared a horrid African fight about the differences between Keith Johnston and himself," Kirk wrote to Bates, 5 March 1880, R.G.S. archives.

lands of the Bunga, and, with an effort that almost ended Thomson's life, finally attained the southern highlands of what is now Tanzania. Everywhere the white man was regarded curiously; chiefs put him on display and villagers marvelled at his habits. Among the Hehe, whose prowess in war was highly regarded, he and his men were regarded with suspicion. But the Hehe eventually let him continue towards the southwest and Thomson, unlike Richard Burton, was positively attracted to these proud and disciplined people.[1] The following narrative indeed contains comparatively lengthy and still valued descriptions of these people and several of their neighbours. It is to be regretted, however, that Thomson failed to take advantage of his foray into this part of Africa to gather fuller collections of historical and ethnological data.

In September 1879, despite a serious bout of fever, Thomson made his way to the shores of Lake Nyasa, tore off his clothes, and gleefully dashed into Africa's third largest body of water. He felt himself appropriately baptized an African traveller; the first part of his mission had thus been accomplished with commendable economy and efficiency. He had justified the trust originally placed in him by the Royal Geographical Society. But the comparatively calm passage that he and his companions had enjoyed may reflect less upon Thomson's undoubted abilities than upon the astuteness of Chuma and Johnston's original preparations. Certainly by concentrating rather singlemindedly upon reaching his first objective intact,

[1] For Burton's comments, see his *The Lake Regions of Central Africa* (London, 1860), I, 239.

Thomson avoided making more than cursory, albeit competent and often very illuminating, observations of his surroundings. He also failed to reflect in any constructive fashion upon the problems of communication that had bulked so large in the instructions originally detailed to Johnston. Without losing a porter or scrapping with a chief, Thomson had rather unpretentiously, but swiftly and commendably—in the honest, workaday fashion that is exemplified by the present narrative—taken his expedition to Lake Nyasa.

Thomson could easily have re-traced his steps without incurring the displeasure of the Society. Instead he wisely decided to cross the intervening plateaux to Lake Tanganyika. He knew that the arm-chair geographers of Europe hoped that "if intercourse could be established between those lakes there would then be an almost unbroken chain of water-communication from the Nile down to the Zambesi, by way of Albert Nyanza, Victoria Nyanza, Tanganyika, and Nyassa."[1] And so he made his way over the hills and across the rivers that separated the two great lakes, visiting in the process the villages of a number of African tribes hitherto unknown to Europe. His traverse of the Nyasa-Tanganyika plateaux preceded by but a few days, however, the similar exploit of James Stewart, C.E., on behalf of the Livingstonia mission of the United Free Church of Scotland. Both camped at Pambete, the village on the southern shores of Lake Tanganyika from which Livingstone (and Chuma) had first sighted the lake in 1867,

[1] Sir Rutherford Alcock, speech of 12 January 1880, in *Proceedings*, II (1880), 133.

their efforts having jointly opened-up a direct route that was subsequently to be much used by missionaries and traders. (More precisely, although Thomson had reached the lake first, Stewart had followed an easier path. It proved the basis for the subsequent cutting of the road between the two lakes that was named after James Stevenson of Glasgow, a merchant who provided the necessary funds.)

Then suffering intensely from malaria and an inflamed tooth, Thomson could with abundant honour have returned with Stewart to Bandawe and thereafter have taken his men by a well-travelled route down the Shire and Zambezi Rivers to the Indian Ocean and Zanzibar. He had completed the mission entrusted to Johnston so many months before. "We may claim," the President of the Society announced, "as probably the most remarkable event in African exploration during the year, the journey [by Thomson] across the previously unknown tract of country separating the two great lakes Nyassa and Tanganyika. . . ."[1] But the riddle of Lake Tanganyika's curiously fluctuating level was still unsolved, the precise nature of the Lukuga outlet continued mysterious, and, having reached the center of the poorly known "dark continent," he understood that it would be rash not to continue onward while the possibility remained real of adding to Europe's knowledge of Africa, and, presumably, to the lustre of his own fame and glory. "Tomorrow," Thomson wrote home modestly in November, 1879, "I start along the Lake North till I reach

[1] Earl Northbrook, "Annual Address on the Progress of Geography," 31 May 1880, *Proceedings*, II (1880), 409–410. Lake Nyasa was originally spelled with a doubled "s".

a good place where I may camp my men and taking about 30 push on for the Lukuga Creek, passing down the river about 30 miles, then striking South through the still unexplored region in that quarter get back to Camp and march for Kilwa which I hope to reach within six months!"[1]

It proved a typically over-optimistic programme and schedule. Nevertheless, after bivouacking the body of his followers near the mouth of the Lofu River and disregarding warnings that the paths beyond were difficult, the mountains inhospitable, food scarce, and the Africans *en route* fierce and ungovernably savage, Thomson and his small band headed north. The following pages contain the only complete account of their vicissitudes along the south-western marches of the lake, their trying passage through Marangu (the country of the Holoholo), and their rapid march through troubled Uguha. And once they reached the banks of the Lukuga River, the resolution of the problem of Lake Tanganyika's wildly-fluctuating level, and answers to queries about its outlet, seemed obvious. Burton and Livingstone knew nothing of any effluents. Verney Lovett Cameron, who carefully sailed around the lake in 1874, had observed that the Lukuga flowed outward. Two years later, when Stanley investigated the same junction of river and lake, he concluded that the Lukuga could function as an outlet only when it was not blocked by papyrus, rank growth, and mud.[2]

[1] Thomson to the Secretary of the Royal Geographical Society, 9 November 1879, R.G.S. archives.

[2] Thomson's comment (below, II, 65) is misleading on the results of these previous investigations. See Cameron, *Across Africa*, (London, 1874), I, 304–307, 312; Henry Morton Stanley, *Through the Dark Continent* (London, 1878), II, 10–11, 44–52.

In view of these earlier investigations, Thomson wanted to know whether the waters of Tanganyika discharged down the Lukuga, whether its vegetation formed a permanent bar, and what relation, if any, the problem of the Lukuga had to the level of the lake. On Christmas day, 1879, he seemingly obtained conclusive results; he needed to make no observations with portable levels or innumerable straws and branches. The Lukuga clearly emptied Tanganyika and, with no obstructions in sight, its waters presumably raced westward to the Lualaba and Congo Rivers and thence into the Atlantic Ocean. Stanley's barrier of mud and sedge had been destroyed and the rapid rush of riverborn flood cascaded through the narrow rock-strewn gorge that both Cameron and Stanley—whose visits occurred during the dry season—had seen dammed. Thomson also sensibly inferred that evaporation usually exceeded or at least balanced the amount of rain that reached Lake Tanganyika during any given year, that this balance, or, he might have added, a period of prolonged drought, caused the level of the lake to fall and the volume of water that flowed down the Lukuga to diminish. As the river became more sluggish, it silted up, debris formed a barrier among the rocks, and the lake, deprived of its outlet, slowly rose until the river overflowed and destroyed the barrier. This was a geographical contribution as important as Thomson makes it out to be in his book. It was subsequently confirmed by numerous, more careful, inquiries. But unbeknown to him, and to his subsequent chagrin, Captain Edward Coode Hore of the London Missionary Society had, six months before, anticipated Thomsons's discoveries.[1]

During the early months of 1880 Thomson attempted to trace the Lukuga to its confluence with the Lualaba River. The second volume of *To the Central African Lakes* contains a full account of the hostility that he and his men encountered in Urua, their ensuing trials, and their ultimate failure to find a route to the headwaters of the Congo River. Instead Thomson beat a hasty retreat to Lake Tanganyika and, together with Hore, sailed south to rejoin his caravan. In the course of this voyage, described hereafter and in Hore's unpublished journal, Thomson visited Karema, an outpost of King Léopold II's African International Association, and later drew heavy Belgian criticism as a result of his own scathing condemnation of Karema and its personnel.[2] Thomson believed that the Belgian and Irish pioneers at Karema had a haughty disregard of the feelings and customs of Africans and, in consequence, understood them to have enraged the usually peaceful local inhabitants. They, in turn, thought Thomson's professedly pacifistic methods quaint and foolish. During the remainder of his voyage south, however, Thomson had good reason to regard his own methods as superior. Twice he might have lost his life if the suspicions of armed Africans had been aroused by antagonistic behaviour. In both cases the Africans mistook Thomson and Hore for Arab slavers.

[1] "The Outlet of Lake Tanganyika," *Proceedings*, II (1880), 41; Edward Coode Hore, *Tanganyika: Eleven Years in Central Africa* (London, 1892), 113, 147.

[2] Hore's Journal is held in the archives of the London Missionary Society. For Thomson's comments, see below, II, 184, 186. They were vigorously challenged by Jerome Becker, in *La vie en Afrique ou trois ans dans l'Afriquc Centrale* (Bruxelles, 1887), II, 115–135.

Thomson returned triumphantly to Chuma and the bulk of his men in April, 1880. Not a man had deserted, none had died, and the camp remained as clean and neat as the explorer had left it. Chuma, Thomson admitted, was a treasure who could not be valued too highly.[1] From him he learned, however, that Thomson's intention to return to Kilwa *via* Ubena and the Rufiji River valley would have to be abandoned. Merere, chief of the Sangu, and Mamle, chief of the Hehe, were at war. Thomson and Chuma therefore decided that the only open route lay northeast-wards through Ufipa to Tabora, and thence homeward along the oft-travelled road to Bagamoyo. Aside from his detour to view Lake Rukwa, the progress of Thomson's journey to the coast needs little comment and is, in any event, amply discussed below. Many travellers had heard of Rukwa, a few had described its general configuration from African reports, but none had ever placed it accurately on a map or reported intelligently upon its character. Thomson therefore decided to deviate from his line of march in order to verify the existence of the lake and to give precision to its outline. Led by the son of a Fipa chief, Thomson thus left most of his men near the modern Sumbawanga and trekked over several mountain ranges into the drainage basin of the lake. Rukwa appeared in the distance, but no matter how many hills he climbed, the lake seemed to appear no more accessible. Unwilling to spare the several days that it might take to descend the escarpment and marshy plain that lay between his view-point and Rukwa's edge, Thomson thus contented himself

[1] FO 84/1574: Thomson to Kirk, 27 May 1880.

with a series of distant observations. The geographical prize, strictly speaking, eluded him, but he did determine that Rukwa drained an area comparatively larger than Lake Tanganyika and that it had no outlet. He also tried to name the lake Leopold, after England's then student prince (later the Duke of Albany), but, possibly because of the possible confusion with Lac Léopold II in the Congo, it never retained the royal designation. And he left the task of ascertaining the precise nature of the lake to subsequent travellers.[1]

"My fondest boast is," Thomson told his Scottish friends after he had successfully completed the remainder of his journey to the coast in record time and had returned home, "not that I have travelled over hundreds of miles hitherto untrodden by the foot of white man, but that I have been able to do so as a Christian and a Scotchman, carrying everywhere goodwill and friendship, finding that a gentle word was more potent than gunpowder, and that it was not necessary, even in Central Africa, to sacrifice the lives of men in order to throw light upon its dark corners."[2] He was less confident of his scientific accomplishments, and the published narrative that follows admittedly includes very few descriptions of the physical setting of his travels, but botanists and zoologists (their reports are contained hereafter as appendices) were reasonably pleased with his

[1] Dr. Emil Kaiser died at the lake in 1882. See his, "Reise von Gonda zum Rikwa-See," *Mittheilungen der Afrikanischen Gesellschaft in Deutschland,* IV (1883), 91–95; David Kerr-Cross ("Geographical Notes on the Country between Lakes Nyassa, Rukwa, and Tanganyika," *The Scottish Geographical Magazine,* VI [1890], 290–292), reached the shores of the lake in 1889.

[2] Quoted in the *Dumfries and Galloway Standard and Advertiser.* 15 September, 1880.

collections. The flora that Thomson took to England and deposited in the Royal Herbarium at Kew Gardens included nearly 200 species, many of which were related to known South African or Ethiopian genera. His sample also included a previously unnoted tree fern to which his name was affixed (*Cyathea Thomsoni*), a new scapigerous *torenia*, and a new *tecoma*. But his conchological specimens were more important: Until Thomson transported them to the British Museum only a few imperfect examples of Africa's lacustrine conchology were known. Of Thomson's forty-seven different varieties of shell life, no less than fifteen were species new to science, Two, including *Limnotrochus Thomsoni*, were new both generically and specifically. Unfortunately, however, his geological observations were of a less important order. He compiled a geological sketch map of eastern Africa and arrived at a number of surprisingly accurate conclusions regarding the composition and configuration of the terrain across which he had trekked, but his analyses derived more from a general appreciation of the countryside, educated guesses based upon sketchy empirical data, and highly developed intuition. He chose not to spend his days in Africa making detailed investigations of a more lasting kind.[1] The role of caravan leader proved too demanding. And it is in that capacity, and as a

[1] For a detailed report on the conchology, see Edgar A. Smith, "On a Collection of Shells from Lake Tanganyika and Nyassa and other Localities in East Africa," *Proceedings of the Scientific Meetings of the Zoological Society of London* (1881), 276–300. For the geology, Joseph Thomson, "Notes on the Geology of East-Central Africa," *Nature*, XXIII (2 December 1880), 102–104; and his "On the Geographical Evolution of the Tanganyika Basin," *Report of the British Association for the Advancement of Science, 1882* (London, 1883), 622–623.

geographical pioneer, that he in fact made his most important contributions to Europe's growing awareness of inner Africa.

Thomson agreed to transform his journals into the present book only reluctantly. With the editorial help of his eldest brother, a literate Presbyterian clergyman, he wrote rapidly and with an almost instinctive flair for simple, direct expression. Sampson, Low, Marston, Searle, and Rivington, of London, whose African series was well known, published it in two volumes early in the summer of 1881. Despite the almost simultaneous appearance in English of Emil Holub, *Seven Years in South Africa* (London, 1881) 2v., and Alexandre Alberto da Rocha de Serpa Pinto (trans. Alfred Elwes), *How I Crossed Africa from the Atlantic to the Indian Ocean Through Unknown Countries* (London, 1881), it quickly passed through several printings. Houghton, Mifflin, & Co. of Boston brought out an American edition later in the same year and Hermann Constenoble of Jena issued a German version early in 1882.

The British reviews were uniformly favourable. It was nonetheless immediately evident that Thomson's unvarnished exposition of a memorable journey lacked the important qualities that had so distinguished the publications of Livingstone, Burton, Stanley, and (Sir) Samuel W. Baker. Thomson's tact and his very equanimity detracted from his memoirs as much as they enhanced his personal stature. His Scottish "canniness", enthusiasm, energy, and tenacity is amply reflected on the following pages, but the intensity of purpose that so marked Livingstone's works, and the vitality of Stanley's, is nowhere mirrored by

Thomson. Where Livingstone and Burton analyze and reflect, Thomson sketches, and passes on. Symptomatically, *To the Central African Lakes* is one of the very few nineteenth century narratives of travel to include no engravings of people or places. Like everything that Thomson ever did (he later went up the Rovuma River for the Sultan of Zanzibar, traversed Masailand, obtained treaties in Northern Nigeria, explored Morocco, and travelled for Cecil Rhodes in what is now Zambia), the following two volumes are interesting, reasonably thorough, and unquestionably competent. They convey Thomson's sense of humour and gay aplomb. They unquestionably deserve once again to be made available. But they are by no stretch of the imagination brilliant; like so much else that Thomson did, they are the product, indeed a very worthy product, of a lively, untrained, and undisciplined intelligence. None so young had ever before withstood so many crises, travelled so far under such difficult circumstances or, in sum, performed so many tasks so well. Very readable, the following volumes belong to an age when young Britons could, and often did, make their mark by performing courageous deeds in foreign parts. They recall the exploits of a prominent explorer who admired Africans, felt at home in their company, and refused to reach his goals by martial means. Recognition for Thomson's accomplishments is long overdue.

R.I.R.

New Grove House,
15 November 1966

KEITH JOHNSTON.

From a Photograph by Mr. Henwood, of Turnham Green.

TO THE

CENTRAL AFRICAN LAKES

AND BACK:

THE NARRATIVE OF THE ROYAL GEOGRAPHICAL SOCIETY'S

EAST CENTRAL AFRICAN EXPEDITION, 1878-80.

BY

JOSEPH THOMSON, F.R.G.S.

In Command of the Expedition.

WITH A SHORT BIOGRAPHICAL NOTICE OF THE LATE MR. KEITH JOHNSTON, PORTRAITS, AND A MAP.

IN TWO VOLUMES.

VOL. I.

London:

SAMPSON LOW, MARSTON, SEARLE, & RIVINGTON.

CROWN BUILDINGS, 188, FLEET STREET.

1881.

[*All rights reserved.*]

PREFACE.

A journey to the tropical regions of Africa has always been a matter of peculiar interest to the English nation, connected as its traditions are with almost the entire work of the exploration of that Continent. Though the subject of African exploration has of late lost a little of its novelty, by the settlement of the most puzzling of its problems, such as those of the Nile and the Congo, it still attracts no inconsiderable share of attention, and any new light is eagerly sought for. In this belief, I venture to lay before the public the simple narrative of the latest effort made by the Royal Geographical Society to unveil some of the mysteries which yet enshroud the "Dark Continent."

Mr. Keith Johnston, who, on account of his well-known and varied abilities, was chosen to lead the

Society's East African Expedition of 1878, succumbed at the very outset of the undertaking; and I, an unskilled youth, who had been selected only as an assistant, found myself unexpectedly in a position of difficulty and responsibility which I felt myself far from being competent to fill. But I remembered I was the countryman of Livingstone, and my Scottish blood would not allow me to retreat till I had performed my duty to the best of my ability.

It was not to be expected that I would bring back the rich harvest of accurate geographical facts which Mr. Johnston would, without doubt, have reaped; neither have I the mature literary and scientific experience which he could have brought to bear on their elucidation and description. Yet the Expedition has been by no means fruitless. The Royal Geographical Society has, through its African Fund Committee, expressed in most flattering terms its approval and gratification at the work achieved, and borne testimony to the fact that the objects of the Expedition have been fully attained. Sir Rutherford Alcock, the energetic and able chairman of that committee, has publicly declared—" I do not know that there has ever been a more successful exploration in Central

Africa, or one more complete in all its parts." The President of the Geographical Section of the British Association Meeting of 1880 has also described it as the most successful and brilliant on record.

In carrying out the objects of the Expedition, an immense area of country has been traversed for the first time. I have had the honour of being the first to reach Lake Nyassa from the north, to journey between Nyassa and Tanganyika, to march along the west side of the latter, and to pass for sixty miles down the Lukuga. Lake Leopold has also been visited for the first time, and some light has been thrown upon a variety of geographical subjects—such as the rivers Ruaha and Uranga, the mountainous region north of Nyassa, and the interesting question relating to the drainage of the Tanganyika.

The Expedition, whose story I now present, has also been unique in many ways. I have to record, neither desertions, deaths (with one exception), plundering by the porters, battles, bloodshed, nor other disasters hitherto supposed to be inevitable adjuncts of African exploration. On the contrary, I cannot express, in too appreciative terms the honesty and faithfulness which characterized my men, and the really genuine character which lies at the bottom of their semi-savage nature. Of the natives likewise I have, for the most part, nothing but good to say. In the majority of places I found them peaceable, and not given to rapine and murder. Rarely did they attempt to throw obstacles in my way. Almost everywhere I was received with genuine hospitality and friendship. The Expedition performed its work with unusual speed; yet never did it come once into warlike collision with the natives, and never had I occasion to fire a single shot at them, either offensively or defensively.

As the result of the Expedition, considerable additions have been made to various branches of science. Several new tribes have been discovered and described, so far as my opportunities for observation and inquiry made that possible. My account of native customs may be thought to be somewhat meagre, and this I admit at once. My travelling experience has convincingly shown me that no one can hope to become genuinely acquainted with African society without a long residence among the people. The traveller passing through the country sees practically very little. Many people will be astonished to learn, that during the fourteen months I was in the interior

I never once saw an African marriage, or the burial of a native, or the ceremonies on such occasions as the birth of a child. My aim has been to describe *only what I myself saw*. Hence the comparative scantiness of anthropological details.

The same may be said about the Natural History. It is almost impossible to form anything like an accurate notion of the fauna of a country through which one is hurriedly marching, burdened with the dreadful incubus of a caravan. With the flora, of course, it is somewhat different, and no inconsiderable additions have been made by my collections to our knowledge of the botany of East Central Africa.

For the first time an attempt has been made to give some general idea of the geological formation of the region of the Great Lakes. Conchology has also received contributions of a very interesting and valuable nature.

These facts I hope afford by themselves sufficient reason for bringing forward in book form a narrative of the Expedition. But there is another still to be mentioned of no less importance, and of even more general interest. I have seen cause to differ from many writers on East Central Africa on such subjects as the prospective trade of that region, and of the value and necessity of railways or roads from a commercial point of view.

Notwithstanding the important facts brought to light by the Expedition, it was with a feeling of dismay that I thought of attempting to put them in a book form. I tormented myself with inward discussions as to the value the public would attach to the opinion or the narrative of an inexperienced youth, however successfully he might have acquitted himself in his enterprise; moreover I had not the slightest experience in literary work, and doubted my ability to produce a readable or presentable work. Impressed by these anxieties, I would have abandoned all idea of proceeding further, but for the encouragement of my good friend Mr. H. W. Bates, who pulled me out of "Slough of Despond," and would not rest satisfied till I had promised to do my best to give the world an account of my doings. He seemed convinced that the narrative of an Expedition so unique in many of its characteristics, could not fail to be of interest to the British public. Mr. Bates' advice has ever been ready at hand, and has been a source of inspiration to me.

The scruples about my literary ability were soon disposed of by my brother, the Rev. J. B. Thomson, of Greenock, who, though burthened with heavy ministerial duties, cordially undertook to correct any literary irregularities in style or expression, and as editor to see the work through the press. To the encouragement, counsel, and aid of these two gentlemen the appearance of this work is mainly due, and I take this opportunity of expressing my obligations.

I have, however, to express my regret that, owing to the necessity of hastening my departure from England to the field of my former exploits, I have not been able to study out and discuss with adequate minuteness many difficult African problems which came under my observation. I have been compelled to finish the work as rapidly as possible, and therefore to confine myself to a large extent to the simple narrative.

Without further apology I send forth this work, relying on the generous and indulgent spirit of the reading public to overlook its many imperfections. It has been conceived amidst fears and uncertainties, and brought forth with much irksome labour, after the energetic and active life I have led for some years past.

One word to prevent misapprehension as to the spelling of some of the proper names, and I have done. The reader will find such words as U-zaramo, M-zaramo, Wa-zaramo, and Ki-zaramo. Each prefix has a definite meaning. U—indicates the country, M—a person of the country, Wa—several persons, or the people of the country, and Ki—the language. There are several exceptions to these rules. Thus Mahenge is a country, and the people are spoken of as "Mahenge," or "the Mahenge," instead of Wa-mahenge, or Wahenge. Similarly Marungu and the Marungu. Keeping this before the mind, some apparent confusion will be avoided. The system of spelling adopted has been the simple and admirable one used by Bishop Steere.

THE LATE MR. KEITH JOHNSTON.

A BOOK which goes forth to the world as a narrative of the East African Expedition sent out by the Royal Geographical Society under the command of Mr. Keith Johnston, would be justly considered imperfect without some prefatory notice of its leader, to whom the chief share of the labour fell in the organization of the Expedition, and who was stricken down nearly at the commencement of the long journey into the interior.

Keith Johnston had achieved fame as a geographer before he entered on his career as an African traveller. He had been specially trained for the profession of geography, and his reputation as an author and cartographer was steadily increasing. He was the only son of Dr. Alexander Keith Johnston, of Edinburgh, the author of the "Physical Atlas," the "Royal Atlas," and other equally well-known and successful geographical works. Young Keith was born on the 24th of November, 1844, and was thus only thirty-four years of age when he died. He received the rudiments of education at school in his native city, but from an early age was carefully instructed in geography by his father; and on leaving school he received a further special training under private tutors. At the age of twenty-two he came to London, and entered the geographical establishment of Mr. Stanford, where he remained upwards of a year, superintending the drawing and engraving of maps. During this period he assisted, among other things, in preparing the "Globe Atlas of Europe," and the series of maps illustrating Murray's "Handbook for Scotland." He then, in July, 1867, went to Germany, chiefly for the purpose of perfecting himself in the language, but also with a view to improving his geographical knowledge and practice. He spent most part of his time at Leipzig, but visited also Berlin and Gotha, at which latter place he had opportunities of observing the German methods of work followed at the famous geographical establishment of the Messrs.

Perthes. The editor of Perthes' "Geographische Mittheilungen," and chief scientific adviser of the Messrs. Perthes, was at that time the celebrated geographer, Dr. Petermann, who, in his youth, had been an *employé* of young Keith's father in Edinburgh; and to the friendly attentions of this able man he was much indebted. The experience and knowledge the young Scotch geographer gained during his residence in Germany, there can be no doubt, had the greatest influence on his mind and character. He not only became a sound German scholar, to the extent that the study of the best scientific literature in that language was a pleasure to him, but he imbibed the German habit of thoroughness and painstaking accuracy in all his work. These characteristics were apparent in all he afterwards did, and they became stronger the older he grew. He had a horror of all "scamped" work, and his latest productions were his best.

On his return to England, in February, 1868, he pursued his profession of geographer under his father at Edinburgh for eighteen months, and then came to London, and took charge of the geographical establishment of Messrs. W. and A. K. Johnston. In

1870 he published his "Lake Regions of Central Africa," illustrated by an original map drawn by himself. This little work attracted the attention of Dr. Livingstone, who was then travelling in the far interior; and in one of his letters the great explorer showed the value attached to the views it contained, alluding humorously to the "geographical acumen of Keith Johnston secundus."

In the summer of 1871 his father died; and in the autumn of the same year he offered himself to the Council of the Royal Geographical Society as a candidate for employment on the expedition then preparing for the search and relief of Dr. Livingstone. Although, yielding to the persuasions of his friends at the time, and withdrawing his application, the desire for active work in exploring some new country had taken such firm hold of him that sedentary employment became irksome, and he resolved to accept the first chance of going abroad that came in his way. Meantime he fulfilled the duties of map-draughtsman and assistant-curator to the Royal Geographical Society, until, in November, 1873, the wished-for opening presented itself, in an offer made to him, on behalf of the Government of

Paraguay, to join, as geographer, a scientific commission for the survey of that country. He was employed in this work about eighteen months, returning to England in May, 1875.

This journey of exploration, in spite of the many privations he and other members of the commission suffered through the impecuniosity of the Paraguayan Government, only served to whet the enthusiastic young geographer's appetite for more work of a similar kind. He settled down for a time to steady work in London, preparing, amongst other things, the "Library Map of Africa," published by Messrs. W. and A. K. Johnston, the volume "Africa," forming part of Stanford's "Compendium of Geography and Travel," and his "Physical, Historical, Political, and Descriptive Geography." But when, in 1878, the African Exploration Fund decided on despatching an Expedition to the head of Lake Nyassa, he offered himself as leader of the Expedition, and was accepted. The rest of the story of his life, which is that of the early days of the Expedition, is recounted in the body of the present work.

In private life Keith Johnston had endeared himself

to his friends by his many excellent qualities. He was singularly unselfish, and considerations of worldly or commercial success in anything he undertook seemed never to enter his mind. It is related of him that when once told that a certain publication of great utility that he projected would never pay, he replied "that was the last thing to think of." He was upright and manly in every thought and deed, and though firm as a rock in carrying out his views of what was right, his manner was always free from all appearance of presumption or self-assertion. If he had a fault it was his exceeding reticence; for it was only to his oldest and most intimate friends that he would occasionally unbend, and speak of himself and his aspirations. At the same time he was fond of all athletic exercises. He took prizes at school for fencing and gymnastics, and was a great walker. For several years, and, in fact, up to the time of his leaving for Africa, he might often have been seen up the river stroking an eight of the Grove Park Rowing Club, of which he was one of the founders; and one year he won his heat in the Thames Challenge Cup at Henley regatta.

As a consequence of this devotion to manly exercises he was of well-knit frame and comely appearance, and seemed always in perfect health and training. All the more surprising was it to the numerous friends who mourned his untimely death, that he should have succumbed so soon to the effects of the African climate.

H. W. B.

CONTENTS.

CHAPTER I.

ZANZIBAR.

PAGE

Origin of the East Central African Expedition — Mr. Keith Johnston is appointed Leader, and I as Geologist and Naturalist—Our instructions—Voyage out—My Visit to Berberah—Arrival at Zanzibar—Dr. Kirk's Hospitality—Society in Zanzibar—The Climate—Recent Sanitary Improvements — Characteristic Sights—Variety of Races—Residence at the Old Consulate — Reception by the Sultan — Celebrities met—Recreations—Syed Bargash 1

CHAPTER II.

PREPARATIONS FOR OUR ENTERPRISE, AND EXCURSION TO USAMBARA.

Our study of Ki-Swahili — Engagement of Chuma—Sketch of the man—Susi and Jacob Wainwright—Selection of Porters — Purchase and Packing of Stores—Common Mistakes in Outfit—We set out for Usambara — Pangani and its Surroundings — A typical African Village—Botanical Character of the Country—Magila and its Mission—Into the Mountains—" The forest primeval "—Native Suspicions—

PAGE

Arboreal Monstrosities—Hendei, its Chief and People —A "Baptism of Fire"—Forest Adventures—Return to Zanzibar — Getting "Seasoned" — The Prickly Heat—Visit of inquiry to Dar-es-Salaam—Shipping our Caravan—Under Weigh 29

CHAPTER III.

DAR-ES-SALAAM TO MKAMBA.

East African Coast Scenery—Dar-es-Salaam; its History and Surroundings—Expectations *versus* Experience —Natural History Notes—The Tropical Twilight—Final Arrangements for the March—The "Characters" of the Caravan—Ready for the Road—Order of Procession — The Start — First Day's Journey — Evening Employments—The Waswahili—Crossing the Mzinga—In Camp; a Dance of Porters—A Ducking—Humours of the men—Our First Sunday on the march—Wazaramo Houses and Fashions—"Casting out Devils" 70

CHAPTER IV.

MKAMBA TO BEHOBEHO.

Harvest time at Mkamba—Native Fête in our honour—Reconnoitring—A Hippo Hunt and its Results—Preparing for a Desert Journey—Punishment of a Porter — An Affectionate Husband — A Miserable March—Desertion of our Guide—The Leaders *hors de combat*—An Unfortunate Decision—The Stillness of Solitude — The Unyamwesi Donkey — Riding *versus* Walking—Johnston's Sufferings—The River Rufiji — A Weary Desert — Sporting Attempts—Snakes — Arrival at Behobeho — Review of the Country passed—The Wazaramo, their Character and Condition—Burton's Conclusions examined; recent Advancement explained—Improvability of the Negro —Export Trade of Uzaramo 105

CHAPTER V.

TO THE MOUNTAINS.

PAGE

Forest and Village of Behobeho—Our Leader fast sinking A Strange Mountain — Death of Johnston — My resolve taken—Onward! — Disquieting Rumours—A Perilous Scare—Conference with the Savages—The Wakhutu — Their Degradation and Climatic Disadvantages—Scenery on the Msegwe—Mgunda—Tribes near Khutu—Khutu House-building—Cutting our way—A Lifeless Region—The River Ruaha—First Voyage of the "Agnes"—Mahenge reached—Baffled by the Porters—Reception at Mkomokero—Examination of the Uranga—Secured for Exhibition—On View—A War Dance—The Mahenge Country, People, Products, and Customs—*En route* once more—Injured Innocence—Rumours of Carter's Elephant Expedition—Through the Lowlands 144

CHAPTER VI.

THE PLATEAU OF INNER AFRICA.

First Part of the Campaign over—Mountaineering—A Life and Death Struggle — Charms of Mountain Scenery—A Terrific Night—The Hunger Ordeal—Heroic Porters—A Race for Supper—The Plateau of Inner Africa—A Stampede—Wahehe Houses—The Country of Uhehe—A Polite Hint—Dress of the Wahehe — Punished for our haste — Educative Attempts — Rebellious Mutterings — A Crisis — Deserted—The Master mastered—Ubena—A Picturesque Chaos — Native Superstitions—Detention at Uhenge — Later History of the Wahehe — Wahehe Customs and Characteristics — Gentleman Savages — Description of the Plateau — A Fête at Uhenge—Released—An Intolerable Entertainment

PAGE

—The Squinting Tribe — Miserable Specimens of Humanity—No Konde Range—Upper Plateau of the Lake Region—Marching in Distress—Nyassa in Sight—A Night Adventure—Headlong Haste—Baptized 199

CHAPTER VII.

BETWEEN NYASSA AND TANGANYIKA.

Pupangada — Search for Lost Porters — The People and their Employments—A Discovery in the Camp —Marriage without a Licence—*En route* for Tanganyika—An Arcadian Scene—Makula's Kindness—Sanitation in Central Africa—A Mistake corrected — Serious Fracas — Leaving the Fleshpots — A Volcanic Region — Delirious — A Word for the Porters—Timidity of the Wakukwe—An Ominous Forecast—The Wanyika—The Highest Point in our Journey—Oiling the Wheels—Stopped by a Petty Chief — Laying a Ghost — Systems of Rationing Caravans—A Pilgrim of Love—Reception by Mlilo —Conflagration in the Camp—Narrow Escapes—Source of Native Iron—Experiences at Sombè's—Tanganyika!—A Proud Moment—Adventure with a Hippo—Sail on the Lake—Retrospective Glances—Geological Notes—The Tribes between the Lakes—History of the Walungu—Walungu Customs . . 260

CENTRAL AFRICA.

CHAPTER I.

ZANZIBAR.

The East-Central African Expedition, whose doings form the main subject of this volume, may be regarded as the outcome of the International Association for the opening up of Central Africa.

The indifference with which the Continental nations had looked upon the Exploration of Africa was suddenly dissipated by the romantic and striking discoveries of our later travellers, Livingstone, Cameron, and Stanley. Annoyed at receiving no share in the honour and glory of this great enterprise, the people of the Continent, with King Leopold of Belgium at their head, resolved to form an International Association to prosecute the work systematically on an extended scale.

England, represented by the Royal Geographical Society, kept aloof from all connexion with that scheme—wisely, as subsequent events have shown. But, jealous of losing its ancient supremacy in the field, it resolved to continue on its own responsibility the work so long and honourably connected with its name. For this end the African Fund Committee was formed, and soon was in a position to organize a new expedition, though on a small scale. There was no difficulty in finding a suitable leader. Mr. Keith Johnston, an athletic young gentleman, already well known in geographical circles, had for some time cast longing eyes towards that land of darkness and danger, Central Africa, and being in every respect preeminently qualified, he was at once entrusted with the command. As Mr. Johnston had gained the character of an enthusiastic and scientific geographer by his explorations in Paraguay and his works on Africa, the appointment met with the cordial approval of all concerned, and a rich harvest of new and trustworthy geographical facts was confidently expected.

In the summer of 1878, just escaped from Edinburgh University, where I had been studying my favourite science of geology, I was wandering somewhat listlessly among the pleasant hills and dales of my native county, Dumfriesshire. Like most young fellows emerging into manhood, I wondered what my lot in life would be, and strove to set before myself some aim to guide my actions. While in this mood of uncertainty, I observed one day a simple paragraph in the newspapers, stating that the Royal Geographical Society were about to despatch an exploring expedition to East-Central Africa, under the command of Mr. Keith Johnston.

That paragraph gave the "turn to the tide of my affairs," and determined my future action.

Though I had never for a moment thought of Africa as the possible field of my future work, yet, like a bright ray of inspiration it immediately struck me that here was ample scope for all my unused energy. Visions of adventure in unexplored lands, and among strange tribes, rose vividly before me. The geology of this great region was also unknown. Might I not with my newly acquired knowledge throw some light on this subject?

In the enthusiasm and excitement aroused by such musings, I there and then sat down and wrote to Mr. Johnston, volunteering to go with him in any capacity, and without other remuneration than my expenses. My offer was favourably received, and on its being laid before the African Fund Committee, with suitable certificates and recommendations, among which I may mention that of Professor Geikie, I had the honour of being accepted as assistant to Mr. Johnston, though I rather blushed to find myself somewhat inappropriately styled "Geologist and Naturalist to the Expedition."

In this manner then, did I become attached to a highly important and arduous work at the rather immature age of twenty, and I must confess that I was about as much surprised by my good fortune as doubtless were most of my friends.

In the instructions issued by the Society, the Expedition was directed to explore the country between Dar-es-Salaam, to the south of Zanzibar, on the mainland, and Lake Nyassa, and to consider the practicability of constructing a road between these places. If our stores were not exhausted on reaching Nyassa, the Expedition was to continue its explorations as far as Lake Tanganyika. Particular attention was to be directed to the still mysterious rivers Uranga and Ruaha, and to the then recently discovered mountains of Konde.

With the slender means at our disposal (1500*l.*) it was thought that the expedition would not be able to accomplish more than the work above specified, though no restrictions were placed on a possible extension of our field of exploration; but, in view of the ruinous cost of some of the later African expeditions, we were strictly enjoined to keep within the sum allowed us. At that time the question of the outlet of Lake Tanganyika was still a matter of doubt, on account of the contrary statements of Cameron and Stanley regarding the Lukuga Creek or River. This problem then dangled before our minds as a tempting bait to attract us forward, and we resolved to make every effort to settle once for all the actual facts of the case.

On the 14th November, 1878, we left London in the B.I.S.N. steamer "Assyria," in which Mr. McKinnon had generously given us a free passage. We had very nasty weather in the Channel, and for three days I endured all the agonies of sea-sickness, to which I am peculiarly liable. A pleasant passage through the Mediterranean compensated, however, for these initial troubles.

We enjoyed a few hours on shore, both at Algiers and Port Said, and in passing down the Red Sea we had the unusual gratification of visiting Jiddah and Hodeidah on the Arabian coast,—two unadulterated eastern towns, with their filth, dogs, and beggars, their gorgeously dressed inhabitants, prison-like houses, and quaint, interesting bazaars.

On the 12th of December we reached Aden, which, despite its character of a "cinder-heap," presents when viewed from the sea, a picturesque scene of no ordinary description. Unfortunately the town and surroundings signally fail to justify the anticipation of approaching voyagers. The shore seems to be eternally swept by the burning blasts of an African sirocco.

Having to wait a fortnight at this place for our steamer to Zanzibar, we resolved to occupy the time by a trip to the African side. Our intention was to visit the great fair of Berberah, famed for the thousands of Somali and Arabs who annually congregate from a vast area of country for the purpose of mutual trade. Ultimately, however, I was left to go myself, as Johnston became apprehensive, from the innumerable delays, that we might not return in time to catch the steamer.

The distance to Berberah is about 150 miles, and the only way of reaching it is by the ordinary open boats or dhows of the Aden, Arab, or Indian merchants. The curiously-shaped specimen of naval architecture in which I took passage, with its low bow and high stern, seemed to encourage the suspicion that it was meditating a descent into the deep; and this suspicion was strengthened by the constant baling required to keep it afloat. The boat was crowded with Arabs, both male and female, and Somali, amongst whom I had to squeeze myself as best I might. A revolting odour arose from the bilge-water, tobacco, passengers, and accumulated filth of years. I had no umbrella, and there was no awning to shelter us from the fierce sun of the Red Sea, which beat down on us with withering effect.

During the passage, which lasted two days, I became sea-sick, and troubled with a disorder still more annoying, considering the manner in which I was packed among the passengers, with scarcely room to move an arm or leg. To add to my troubles, I was not long in discovering that there were other animated beings besides Arabs and Somali on board. Unaccustomed as I am to unpinned insect collections, I felt a creeping sensation come over me as I became aware that certain unmentionable insects had migrated from the Arabs to myself. To say that I grew uneasy and fidgetty, would be but a mild statement of my feelings.

One could not but admire the composure of the Arabs in the face of exceedingly lively external conditions. They calmly smoked, slept, and said their prayers, without a hitch at their garments. The reader will easily understand that with such an accumulation of horrors, I was passing through a very trying ordeal in my apprenticeship as a traveller, and that I felt considerably relieved when at last we reached Berberah.

On entering the partially-enclosed piece of water which forms the harbour, we notice on our right hand several well-built houses. These are the barracks, hospital, and residences belonging to the government of Egypt, who have lately taken possession of the surrounding country, and with much energy built a number of capital houses, brought water in iron pipes from the neighbouring mountains, constructed an iron pier, and otherwise tried to make something of a perfectly useless desert.

At the inner end of the harbour is the collection of mud huts used by the Arab and Indian merchants, surrounded by a vast miscellaneous assemblage of dome-shaped booths, formed of boughs and hides, and occupied by the Somali who flock to the fair. Great herds of donkeys, cattle, sheep, goats, and camels are seen in various places, doing their best to still the cravings of hunger on the scanty tufts of vegetation which by careful search are to be occasionally found. To the south extends as far as the eye can reach a dreary sandy desert, which to the west is bounded by a range of high mountains.

An immense hubbub fills the air; shouting, yelling, and quarrelling seem to be the order of the day; and every now and then soldiers are seen rushing hither and thither to stop some scene of blood and murder, such tragedies being of hourly occurrence in spite of all precautions.

The Somali, with their wild faces and frizzly shocks of hair, toga-like dress, cruelly barbed spears and arrows, are everywhere moving about in crowds, and look sufficiently dangerous to make us somewhat chary of mixing with them. On landing, I felt it rather trying to my nerves to be suddenly surrounded by such a savage-looking crew, dancing about, shouting and brandishing their arms as if they had gone mad.

I had rather a strange and varied experience of different races in my first day's stay in Africa. The programme opened with a morning meal in the house of an Arab, on dates, raisins, and eggs. At midday I took dinner with a Greek who had been a cook's mate on board a ship. He contrived to present me with a very savoury mess, which had to be eaten without knives, forks, or spoons, these being unknown luxuries at Berberah. In the afternoon, I had further refreshment with a Hindu merchant, who, looking upon himself as a British subject, felt bound to take me under his care and protection. His bright little nephew acted as my guide. In the evening I finished off the day's round of feasting and visiting by having supper with the Egyptian officers stationed at Berberah. A Maltese, who acted as harbour-master and lighthouse keeper, the Greek trader, and the Hindu boy, were also of the company. Before the day was over, I had made the acquaintance of a wonderful variety of dishes, and had learned that after all spoons, forks, and knives were not entirely indispensable.

At night, owing to the constant state of riot in town and the frequent murders, I was accommodated in the fort by order of the pasha in command.

The next morning I had an interview with this important official. He received me very pleasantly, and gave me an escort of cavalry to conduct me across the desert to the mountain ranges, twelve miles off, whose geological formation I desired to examine. Without such an escort it would have been unsafe to venture so far, owing to the reckless bands of Somali roving about the district. This desert ride I enjoyed very much.

On the following day I was so ill and weak that I could not move about, and as a particular star seemed to harbour some evil design, it was thought advisable not to set sail. On the succeeding evening, the wicked star having apparently given up its evil intentions, we shook out our huge lateen sail to the breeze, arriving at Aden on the next day.

On the 28th of December we left Aden, and, having a strong south-west monsoon on our stern, we sped swiftly through the water towards our destination. In passing near the dangerous coast of Cape Guardafui, we attempted to relieve a stranded steamer, but failed, and had to leave it to its fate. We passed another ship on shore at Pemba Island.

About sunset on the 5th of January, 1879, we arrived opposite the north end of Zanzibar, which, with its evergreen mantle and waving cocoa-nut trees, was a refreshing sight after the dreary barrenness of the country on either side of the Suez Canal, the burnt-up mountains enclosing the Red Sea, and the glaring sands of Berberah. The sunset was exquisitely striking, and lent an additional charm to the transformation scene from the intense glare of day to the cool starry night. Enchanted with the scene we stood silent, watching the rapidly changing tints and shades, till our attention was attracted by the lights of Zanzibar in the distance.

It was dark when we neared the narrow channel which forms the passage to the harbour.

The Captain was complacently making a quiet joke over the "lame ducks" we had passed at Guardafui and Pemba, when suddenly he was stopped by a bump, a shouting of men, and the reversal of the engines making the water boil at our stern. We had ourselves got into the predicament of the "lame ducks," having run on a sand-bank, through mistaking a dhow for a buoy. The Europeans in the town, who were anxiously watching our approach, got a considerable fright about the safety of their letters, but fortunately we were not detained long. The tide was rising, and in an hour we were floated off. Entering the harbour we cast anchor among the twinkling lights of the ships, and in a perfect forest of dhow masts.

Early next morning, as soon as we had despatched breakfast, we proceeded to call upon our accomplished and well known Consul-General, Dr. Kirk, himself a well-known African traveller.

We found him almost buried in the midst of his newly arrived letters and newspapers. On introducing ourselves we met with such a hearty welcome as at once put us quite at our ease. Mrs. Kirk, who presently appeared, also charmed us by her graceful and hospitable reception. We were forthwith invited to take up our residence with them till the house known as the Old Consulate—at that time vacant—was got ready for us. Before returning to the ship we met four children of Mrs. Kirk's, and were much surprised at their healthy appearance. They looked as vigorous and lively as if they had never been out of England, and yet they had never left Zanzibar, where they were born.

During our voyage out, filled with melancholy ideas regarding the deadliness of the climate, we had begun at an early stage to "fix" ourselves up with daily doses of quinine, in anticipation of coming fevers; and now to our amazement here were a lady and gentleman who had spent about fourteen years in Zanzibar, had brought up a large family of healthy children, and were still as full of energy and vigour as if they had been living in the most bracing of climates. We returned to our steamer, the "Punjab," with more cheerful and encouraging views of the climate.

All our impedimenta were hauled out of the hold, and having escaped the trouble of a custom-house examination through the influence of Dr. Kirk, we soon had everything safely stored in the Consulate, where we took up our abode.

The day was spent talking over East-African affairs, which, of course, were intensely interesting to us. Views expressed by one so well qualified to speak as Dr. Kirk, we naturally regarded as of the highest value, and treasured them accordingly.

In the evening we had a walk in the outskirts of the town, to the famous Mnazi Mmoja, or "one cocoa-nut tree," where, among the old Arab tombs, with the roar of the ocean at their feet, the European children are accustomed to play, and where the adults enjoy their favourite evening promenade. The view, looking south across the Malagash inlet, when the tide is in, is one of supreme beauty. In front is a waving mass of vegetation, formed of light-green cocoa-nuts, rising lean and scraggy above the other trees, dense masses of mango with their dark green colour, bananas with their huge ragged leaves, and oranges and lemons to fill in the gaps, and give variety to the picture.

Here and there Indian-shaped huts, with their low

mud-and-wattle walls and immense thatched roofs, appear among the trees. On the left lies the town, with its eastern architecture, and its incongruous huge English church, and numerous consular flag-staffs. On the right is the ridge of the "one cocoa-nut tree," over which are seen the distant bluff of the Mbwenni (Dr. Kirk's country house), and the blue waters of the Indian Ocean. At our feet lies the Malagash inlet, which at high tide adds to the scene all the charms of a placid lake. Everywhere are seen gorgeously dressed Arabs, riding on horseback, or on Muscat donkeys famed for their paces. Negroes dressed in white shirts, or women with more highly coloured stuffs carrying huge pots of water on their heads. With these are mixed Hindus, Banyans, and Parsees, and not unfrequently a European lady or gentleman goes riding or walking past.

The day after our arrival, according to Zanzibar custom we proceeded to visit the different members of the European community—men, according to Stanley writing a few years ago, who "have dwindled into pallid phantoms of their kind, into hypochondriacal invalids."

Introduced by Dr. Kirk, we first called upon the Consuls of the different nations represented officially. We found the French Consul a studious gentleman of retired habits, and surrounded with a cultured atmosphere, such as could hardly have been expected so far from the centres of civilization. The German official, who was only what has been called a "trading Consul,"—that is to say one who combines trading with his consular work,—presented a marked contrast to the Frenchman. Tall, muscular, and fair-haired, he illustrated the national physique in a very distinctive manner. We certainly failed to find a "phantom" here. Calling next on the American Consul, we met a typical Yankee in features, accent, and activity. We then made the round of the merchants, from whom we received a most hearty and generous reception.

In one short day we became completely disabused of our preconceived notions of the people and place. We came out with the idea, derived from our knowledge of similar parts, that Zanzibar was a perfect pesthouse of fever, where little society, and that not of the best, was to be had. We found, to our surprise, that the Europeans were a most select company of gentlemen, full of health and spirits, and that etiquette and fashion were as dominant as in England—a highly desirable circumstance in such small communities in out-of-the-way places, where one is so apt to discard the proprieties and fall into a negligent free-and-easy style of life. Polo and cricket, with lawn tennis, formed the favourite games, plainly showing there was no lack of energy, while a long walk or ride into the country in the pleasant evenings was a matters of course.

That many-sided traveller, Captain Burton, has

made the remark that, "he was astonished at the inhumanity of English husbands, who when they want to get rid of their wives resort to brutal means, when they might easily get rid of them by natural means, if they but gave them a trip for a few months to Zanzibar." In the face of such a statement, it is rather strange to find that several husbands have taken their wives out; but, if their purpose was of the tragic nature referred to by Burton, it has been strangely ineffectual, for, with feminine perverseness, the ladies from Europe seem all resolved to enjoy a reasonable term of life. The husbands, however, to do them justice, have, under the circumstances, become wonderfully reconciled to the inevitable. Indeed, having once settled with their families in this much maligned spot, they all seemed to have adopted the heroic resolution to live happy ever after.

In making our calls we did not neglect to visit that genial and laborious gentleman, Bishop Steere, whose life for the last sixteen years in East Africa, presents a record of travel, and of literary, philological, and missionary work, so vast and varied in character, as at once to dissipate the notion that the climate of East Africa is destructive of all energy.

The wonderful activity of Dr. Kirk, through so many years of important consular employment—the hard work of the merchants, and the athletic sports in which the Europeans constantly engage, all combine to show that Zanzibar is worthy of a better name, as a residence for Europeans, than some writers have given it.

Bishop Steere has fixed the headquarters of the Universities' Mission (to which he belongs), beside the old Slave Market and on the edge of the Malagash inlet. Of this latter, Stanley says, that "a noxious malaria is exhaled from it," and implies that the so-called unhealthiness of the town is due to this. Yet, strange to say, the missionaries, including several ladies, declare that their residence is much more free from fevers than any other part of the town; indeed, their agents stationed on the mainland come to this "malaria place" to restore their health, when they have broken down through disease.

A vast change, however, has come over the climate of Zanzibar within the latter half of this century. The rainfall, through some cause, has diminished by more than one half since Burton's visit in 1857. At that time it was estimated at twelve feet, though the data were imperfect; and now it has fallen as low as four-and-a-half feet, according to the observations of Dr. Robb, the Consulate physician, who kept systematic meteorological observations while in that position. The diminished rainfall, and the vast improvement in the health of the place, may be mainly ascribed to the clearing away of the mangroves and of the dense vegetation which formerly covered the island, the substitution of clove and cocoa-nut plantations, and the greater extension of cultivation.

There have also been considerable sanitary improvements of late years. Dead animals, filth, and garbage of all kinds, are no longer allowed to be thrown out into the streets, to rot and fester in the sun; and dead slaves are at least placed beneath the soil, instead of being thrown upon the beach.

The town is singularly free from dogs and beggars, those pests of Eastern towns. There is an air of cheerfulness and comfort pervading all classes of society which would be unusual anywhere. But then this is that ideal land where a man can live in abundance at from fourpence to sixpence per day. No starved and ill-used slaves are to be seen, for on cases of inhumanity being reported to the Sultan, the sufferers are at once set free, and made safe from the brutality of their masters. Indeed, this class seem to have a remarkably easy time of it, and have ten times more real liberty than thousands of our clerks and shop-girls. They are commonly allowed to engage themselves out to work; and no caravan ever goes up the country without a number of slaves among the porters. A jolly hearted crew they are, whose motto in life is, "beer, women, and ease."

A word further with regard to the general appearance of Zanzibar. It possesses all the architectural characteristics of an ordinary Eastern town, with this exception, that there is a marked absence of minarets and mosque towers. The square prison-like houses, devoid of ornament, and all white-washed, present little variety. They form narrow crooked lanes, whose only recommendation is their shadiness during the greater part of the day. The town stands upon a jutting point of land called Shangani, which is almost an island when the tide is in. There is nothing European in the scene, with the exception of the ships in the harbour and an occasional white man in the streets.

In other respects it may be said to be a harmonious mingling of materials derived from Arabia, India, and Africa. The better houses are Arabian in their architecture, while the poorer take their style from India. From the latter place the most prominent of the fruit-trees which surround the town are also derived.

The only "sights" in Zanzibar are the Fort, where all criminals are confined (and as the Sultan does not believe in capital punishment, he simply puts the worst criminals into the deepest dungeons, where they soon disappear from this earth); the Palace, consisting of the Sultan's public rooms, where he spends the day and does his official work; and the harem, a barn-like building, connected with the former by a covered passage. The visitor will probably also have his attention attracted to Bishop Steere's grand church—a striking illustration of the misapplication of money and energy, which might have been so much more usefully directed into other channels.

The Arabs are the dominant race, and form the

upper and ruling class. They have not, however, the usual conservative character of Arabs, but are rather liberal-minded, and open to improvement and progress. The Banyans and Hindus form the middle class. They have possession of the trade and purse-strings of the country. The Sultan dares not trust an Arab in anything in which the revenues are concerned; consequently the Banyans have all the custom-houses in their hands, and many of them are extremely rich. The negroes form the mass of the population, and are the working class.

Each nationality has its separate quarter in the town. Pass along some lanes, and but for the houses you might imagine yourself to be in India; for here sits the Banyan, Hindu, or Parsee, each in his characteristic costume, surrounded with his curious collection of articles of trade. Wander towards the palace, and you leave the business thoroughfares of India behind, and enter Arabia in a twinkling. Arabs, well dressed, voluminously turbaned, but withal grave and aristocratic in bearing, meet you at every step with a courteous "How d'ye do?" The surroundings harmonize in every particular. Here is a gaily decorated Muscat donkey, and there a veiled Arab woman.

We continue our march, and from Arabia we seem suddenly to enter a purely African scene. We stand in a market-place, with all the ordinary products lying in heaps in the open air, or sheltered under booths, with attendant negroes. There is a gabbling crowd of natives in all degrees of undress—the wild-looking Somali from the far north of the Sultan's dominion, Wanyamwesi porters waiting for a return caravan to Unyanyembe, representatives of the coast tribes, and, finally, Waswahili slaves and freed men from every tribe comprised within the area of the Great Lakes region. These latter wear cool-looking, pure white shirts and caps—the women, however, glorying in the most gorgeously coloured of handkerchief stuff, which they wind not ungracefully about their figures.

On the 20th of January we took up our residence in the Old Consulate, so called from having been formerly the residence of the British Consul, and where our national flag yet flies in the breeze. This house is occupied by the Vice-Consul, Mr. Holmwood. At the time of our visit this gentleman was in England, so we began housekeeping by ourselves. We bought, under Mrs. Kirk's direction, such crockery and cooking utensils as our simple wants suggested, while, through the indulgent kindness of the same lady, we were supplied with chairs, tables, and a sofa from the Consulate. We of course used our own camp-bedsteads and Austrian blankets for our beds.

A few days after our entry into the Old Consulate our reception by the Sultan occurred. Early in the morning, with inward groans and profuse perspiration, we laid aside our light drill clothes and donned the orthodox evening dress. The ceremony was to take place at nine o'clock; so a little before that time we went over to Dr. Kirk, who was to introduce us. We found him in his consular uniform, with sword and military helmet. Just as we were about to start, much to our dismay, a message came from the Sultan putting off our reception for half an hour. This might seem a small matter, but it was not so, as in the hot black suit we perspired most excessively, though meanwhile I took the precaution of seeking out the coldest draught in the room. At last we were informed that everything was ready for the important ceremony.

On our way to the Palace we were preceded by two fancifully-dressed consular servants with canes in their hands, and followed by two others similarly got up. Some distance from the Palace we found the streets lined with Beluch soldiers on both sides. They were a most villanous-looking set of cut-throats, with features of that irregular and broken type only to be found in the lowest grades of barbarian society. They were in all stages of dirt and rags, and their arms seemed to have been procured by the spoliation of an antiquarian museum. A more curious and ancient collection could hardly be conceived, and no one was armed like his neighbour. There seemed to be literally no discipline among them. I may add, however, that these troops have, since the time of which I write, been disbanded. Near the Palace the line was taken up by his Highness's new soldiers, who with their neat white

European uniform really deserve all praise for their appearance. They are all negroes, both officers and men, and are armed with swords and Snider rifles. They have been admirably drilled and organized by an English navy officer named Mathews. On our approach we were received with a military salute, and at the same time the excellent brass band of the corps struck up "God save the Queen." Not having anticipated this grand reception, I have no doubt I unconsciously struck an attitude, and tried to look as though I considered it all a matter of course.

On arriving in front of the Palace, with the soldiers massed round us, the strains of "God save the Queen" were exchanged for the Sultan's national hymn, as a plainly-dressed but pleasant-looking Arab stepped from the doorway into the square, and, advancing towards us, shook hands with Dr. Kirk, uttering the Swahili "Yambo?" or "How do you do?" Kirk immediately introduced us, and we then for the first time became aware that it was the Sultan who had thus come forth to meet us.

These preliminaries over, his Highness motioned to us to enter the palace, which we did, he following behind. We passed through the entrance-hall—lined with Persian mercenaries, who then formed the Sultan's body-guard, though since displaced—up a flight of steps, and along a corridor crowded with Arabs in most brilliant colours, and wearing expensive gold-hilted swords and knives, till we finally reached the reception-room. The walls were covered with huge mirrors, reflecting our persons on every side, while half a dozen niches had as many clocks and telescopes, of which his Highness seems to be particularly fond. The ceiling was loaded with crystal candelabra, and the floor was covered with a rich carpet. Along both sides were gilt chairs with crimson-covered seats, on which the Arab courtiers disposed themselves. At the head of the room was the Sultan's arm-chair, where he seated himself, inviting us to occupy chairs beside him. There were the customary inquiries about our health and voyage, spoken in the Swahili language, through an interpreter, though after a few sentences his services were dispensed with, and Kirk spoke directly to the Sultan. Then in came sherbet in crystal cups, followed by spoonfuls of coffee, also in crystal cups bound with gold, and with small gold spoons.

This customary tasting over, we had quite a long and pleasant conversation about our proposed expedition. The Sultan showod an intelligent interest in everything, and promised to assist us as far as lay in his power. After nearly an hour's stay we retired, the Sultan following us down to the streets. The soldiers saluted; the band played; we shook hands, and with a "Kwaheri," or farewell, we returned to the Consulate, pleased with our reception, and yet glad to get released from all the hot-house agonies of an evening dress.

We found Zanzibar altogether a most pleasant

tropical residence, and though the temperature is exceedingly high and uniform throughout the whole year, yet the monsoons, which blow from either north or south continuously, temper the heat, and rob it of all disagreeable characteristics.

There was no lack of excitement and varied employments and amusements. The monthly mail gave sufficient exciting matter to keep every one on the tip-toe of expectation for at least a week before it arrived. During that time nothing of minor importance could be attended to, the usual saying being that, "It would be done when the mail had gone." Then during the few days' stay of the steamer, no one has time to speak to his neighbour. Letters have to be read and the newspapers examined. The greater part of the month's work is compressed into the mail week, and any one who attempted to disturb the European at that time would certainly run the risk of being snubbed.

The passengers also who arrived at Zanzibar, or were passing to or from the Cape, were great objects of interest. There was always some one turning up —a missionary, a traveller, or a trader. At Zanzibar we became acquainted in this way with many celebrities, such as Stanley preparing for his Congo mission, and Major Pinto hastening home from the field of his exploits. Forbes among newspaper correspondents, and Lord W. Beresford among military men, on their way to the Cape. There also we met missionaries like Wakefield, who had devoted their whole life to the enlightenment of the East African. Surveying and cruising ships frequently brought their compliment of officers.

Among the prospective travellers whom we met, the most notable of all was a Belgian, full of infinite graces and smiles, charmingly dressed, and quite irresistible with the ladies, especially when he let forth his soul in passionate song. He was further marked by his profound ignorance, both of the work and the country before him. But that was nothing! As he said, his king had need of him in his philanthropic and enlightened designs for the opening up of Central Africa. His Majesty had said, "Go to Zanzibar, and organize a caravan to travel in the interior." That was sufficient for him, and without considering such trivialities as whether he was fit for the duty or not, he had at once set off. This same gentleman, after getting a few miles from the coast, "fell sick unto death" as he thought. Then the truth flashed on him that he was not suited for the work, and he precipitately returned to the coast, where he soon roused the admiration of the Europeans by his astonishing adventures among wild animals, and savage cannibal tribes, which, though so near the coast, he has had the merit of first discovering. The game was so abundant and tame that all he required to do, to enjoy a good day's sport, was to be carried in a hammock by his men, when without rising he could bring down his victim with unerring aim! However, the climate he found did not agree with him, so he returned to Brussels, to enliven his comrades there.

Besides having such characters as the above to study, we had ample amusement in our odd hours to keep the time from hanging heavy on us—such as evening walks or rides through charming orange-groves, and games on particular days of the week. And then what could be more delightful than to stroll on band-days down the promenade, to sit in the Square, on chairs specially provided by the Sultan for the convenience of the Europeans, and listen to wonderfully well played operatic selections and to study the marvellous collection of divers races in their infinite variety of costumes? These concerts always took place under the shadow of the palace. In front, dotted over with English and native shipping, lay the harbour, with its sheltering islands. On the left stood a tall clock-tower, and and on the right rose the Sultan's barn-like harem, with attached menagerie—the whole enclosing a Square laid with concrete.

On other days the troops are reviewed in this space, and their fine appearance and well-performed evolutions are certainly worth seeing. The Sultan is also introducing Indian cavalry mercenaries, to further add to his glory. He keeps several carriages, which are proving themselves potent instruments of reform, inasmuch as they are compelling him to broaden the streets, and lay them out in a straight line. These carriages he is by no means loth to lend out to the consuls, when their ladies so desire it. Indeed the stables seem to exist mainly for the benefit of tradesmen, officers of the different ships, and visitors to Zanzibar.

The Sultan himself is a most peculiar and interesting character. He is strict in all the observances of his religion, and yet liberal-minded to an unusual degree. Take as an example the following. On Bishop Steere's new church approaching completion, which, be it remembered, is placed as a memorial of the forcible stoppage of the slave-trade in Zanzibar, he sent a book with a list of clocks varying in price up to several hundreds of pounds, and told the Bishop to select whichever he pleased, as he (the Sultan) desired to present one for the new church tower. Surely a remarkable act of liberality in one who belongs to the most exclusive and fanatical religious sect in existence! Many in our supposed-to-be liberal-minded nation might take a lesson from this follower of the Prophet.

Syed Bargash, though generous and genial, is yet a keen man of business, who has always an eye to the main chance. He is now proposing to run steamers on his own account between Zanzibar and Jiddah, to convey pilgrims, who would of course prefer a ship under Mohammedan officers and crew, though vastly less safe than one in the hands of infidel Europeans. There can be no doubt that this will be a paying venture, if they do not wreck the steamers on their first voyage—an event by no means improbable.

Strange to say, the Sultan contrives to keep out of debt, though not unfrequently rushing into unprofitable speculations. He betrays an intense curiosity with regard to everything European, a trait probably originating in his visit to England; and his greatest pleasure seems to be watching the ways of the foreigners from his palace window. Not unfrequently, however, he makes some sad misinterpretation of what he sees; as when he, on one occasion, mistook the somewhat eccentric movements of a very worthy missionary for drunkenness, and sent to Dr. Kirk to know whether he should not in such case have been lodged in jail. In these social investigations he employs a powerful telescope, and thus often becomes acquainted with facts not intended for his eye. He has more than once revealed knowledge of a kind and extent which has rather unpleasantly taken some Europeans by surprise.

CHAPTER II.

PREPARATIONS FOR OUR ENTERPRISE, AND EXCURSION TO USAMBARA.

As soon as we had got comfortably settled down in the cool and roomy precincts of the Old Consulate, we set to work systematically with our preparations. One of the most important of these was the study of Ki-swahili—a language not only spoken by the natives at the coast, but so well known in the interior as to enable the traveller who can speak it to pass almost from one side of Africa to the other. If by any chance he lights upon a tribe, no member of which can speak the Swahili language, he is almost certain to have among his followers one who can speak the dialect of that particular tribe.

One marked characteristic of the Swahili language is its peculiar defining prefixes. This same feature distinguishes all the tribal dialects in Africa, south of the fifth degree of north latitude. Indeed it is one of the principal links connecting the various parts of the great Bantu race of negroes comprised within that area.

The Swahili is soft and melodious. The prevalence of liquids and vowels in it is very noticeable. Almost every word ends with a vowel. Even words adopted from an outside source, and naturally ending with a consonant, have almost invariably a vowel added, as spoonè for spoon.

It is almost impossible to exaggerate the advantage to the traveller of a personal knowledge of this language. If he has even a rough acquaintance with it he can speak to his men directly, and ask questions which a native translator would almost certainly bungle; and what is of quite as much consequence, he can draw closer the bond of sympathy between himself and his followers by having a kindly chat with them. If he is ignorant of their speech, he is of course practically separated from them.

To the acquisition of this language then we devoted our mornings. In our studies we were materially assisted by Bishop Steere, who is universally recognized as the best of Swahili scholars. Mrs. Kirk, who took great interest in all our schemes, also lent invaluable aid, by joining us in Swahili conversation when we took our weekly excursion to Mbwenni, where Dr. Kirk varied his consular duties by vigorously working in his garden.

The formation of our caravan was, of course, a matter of all-absorbing interest to us. Our first important step in that direction was the engagement of Chuma as our chief headman.

To the British public the name and deeds of Chuma must still be very familiar. His long and faithful service with Livingstone in his last days, and his success in the chivalrous, though hazardous, undertaking of bringing his body to the coast, cannot yet be forgotten.

It will not be out of place here to say a word about his capabilities and character. The vague expression of "the Expedition," "we," or the more egoistical "I," is very apt to swallow up a subordinate's individuality and valuable services, throwing him into the background, where he is more than likely to be forgotten. To obviate such a fate for the invaluable headman of the East Central African Expedition, let me anticipate the thread of my narrative while I bring him specially under the reader's notice.

Among the guild of Zanzibar porters there is certainly none to equal Chuma as a caravan leader, especially for white men. His long experience under Livingstone as an interpreter of the geographical questions so necessary to be asked, gave him a very fair notion regarding these things, so that he is able at once to pick up a European's meaning when an ordinary native would only look at him in blank perplexity. He is well acquainted with English, and about a dozen native dialects. For eloquence in speech-making he is unrivalled among his fellows, and it is quite a treat to watch his animated manner and natural gestures. He has all the natural requirements of an orator, and his speeches seldom fail in their intended effect. Whenever such were required, whether to the porters or in the presence of native potentates, I left to him the task, knowing that from his fluency and fervid eloquence, and his knowledge of what would most impress or rouse the native mind, the effect would be far greater than could be produced by any such attempt of mine. Full of anecdote, and fun, and jollity, he was an immense favourite with the men, and yet he preserved such an authority over them that no one presumed to disobey his orders. If any one was rash enough to do so, woe betide the offender! Chuma went straight at him; and though not tall or muscular himself, he speedily humbled the strongest.

The value of a headman possessing such qualities and attainments will be better understood when we consider that the education of the white traveller is entirely in his hands for the first few months. The headman may practically do what he likes with his pupil, and if not faithful and honest, he may impose on him at every hand. In all cases the success of the first part of an African journey must depend upon the native headman.

With all his good qualities, however, Chuma was by no means immaculate or faultless. I soon found him possessed of that vice of all half-civilized negroes—unveracity—a vice which is singularly developed among the natives of East-Central Africa. His off-hand statements required to be accepted with judicious reserve. Lies came natural to him, not indeed from any premeditated purpose, or from desire of gaining profit or pleasure to himself, but simply because they seemed to be always nearer his tongue than the truth. Yet it was almost impossible to catch him tripping—a fact which often made matters extremely exasperating, when we knew that, however plausible his story, it was untrue.

Chuma was also extremely fond of acting the big man, and right well he could do it. To keep up his dignity he deemed it necessary to be somewhat lavish in his expenditure, so that we required to be continually on the look-out, and to keep a firm hand upon him to check his extravagance.

These and other characteristics demanded a stern application of the strictest rules of discipline. But when thus kept within bounds Chuma was an invaluable helper in our enterprise.

The names of Susi and Jacob Wainwright have been associated in the public memory with that of Chuma, and perhaps it may interest some of my readers to hear, in passing, something concerning the recent history of these two former comrades of our headman.

On inquiry we found that Susi had fallen into very bad drinking habits, and was in a state of destitution through his debaucheries. He was, perhaps, even a more able man than Chuma in some respects, and but for his prominent failing, he might have been at least equally successful. He was very desirous of joining our caravan, but, considering that he had always been rather above than under Chuma in his previous engagements, we thought it would not be prudent to have them both, and so declined his services. He is, I believe, at present up the Congo with Stanley.

Jacob Wainwright we also found to have fallen considerably. After his return from England he got an excellent situation with some missionaries on the mainland, but became so impudent and forward that they were compelled to dismiss him. He was in the habit of twitting his European masters with the fact that they had never, like him, had the honour of being presented to her Majesty Queen Victoria. His airs and arrogance in consequence of this honour became quite intolerable, and there was nothing for it but to part with him. When last I heard of him he was acting as a door-porter to one of the Zanzibar traders.

So much for Chuma and his former comrades. We return to the subject of our preparations.

With the assistance of Chuma, and the headmen of Dr. Kirk and Bishop Steere, a most select company of porters was got together. At the time of which I write, the task of forming a good caravan was one of peculiar difficulty. The Sultan, to make up his army, was pressing all idlers into the ranks. Consequently all the nondescript villains and scamps of the town were eager to get engaged as porters, to escape such a fate. It will thus be seen that the utmost caution and discrimination were necessary. Fortunately our subsequent experience of the men chosen was thoroughly satisfactory. Besides the porters, five more headmen were selected, to act as staff officers or aides-de-camp under Chuma.

The next necessary piece of work was the selection of cloths, beads, and wire, and the determination of the different kinds and relative quantities of each to be taken. These were considerations of vital moment, and demanded a good deal of careful study.

From an inquiry among natives and Arabs, we soon learned that fashion was as dominant among Central African tribes as among the *belles* of Paris or London. Each tribe must have its own particular class of cotton, and its own chosen tint, colour, and size among beads. The absence of the required article at any particular point, might mean nothing less than disaster and failure to the expedition, as the people will have nothing but the cloth or bead that happens to be in fashion. Everything else is of no value, and will hardly be accepted as a present. Worse still, the fashions are just as changeable there as here. In one year a tribe goes mad for a particular bead; but the trader having supplied himself with the fashionable article, according to latest news, might, if his journey was long, arrive to find the fashion changed, and his stock just so much unmarketable rubbish.

We had an illustration of this in our own case. Depending upon Chuma's knowledge of the fashion prevalent about Tanganyika, when he was there with Livingstone, we laid in a considerable supply of beads of the required size, composition, and colour. When we arrived at the lake, we found beads of all kinds ignored, and coloured cloths in demand. The beads which we had laboriously transported so far proved utterly useless.

Having given anxious regard to these considerations, and drawn out a carefully studied list of our requirements, we proceeded to the shop of a Hindu named Essa-lila, one of the most extensive merchants in Zanzibar. In a few hours we had spent nearly 500*l*. in cloth, beads, and wire, and on the same day our purchases were brought over to the Old Consulate, to be put up in bales of the required size.

This operation of packing was the occasion of a very animated scene. Perhaps the reader may be curious to take a peep in upon it. Round the sides of a large room were the different cloths—hundreds of yards of the flimsy satini, the worthless English cotton; thousands of yards of merikani, the strong and durable cotton of America, which is fast hustling England out of the African market; piles of various unsubstantial but highly-coloured cottons from India; with gold and silk cloths for chiefs, and gorgeous things innumerable. These different cloths were first folded into the required shape, and passed on to those men who were making up the contents of the bales. To obviate the necessity of opening more than one bale at any particular place, each was made up of mixed cloths. Cotton, white and blue, formed of course the major portion of each; there were so many coloured cloths for special purposes or rewards, and finally one or two more expensive kinds for presents to the bigger chiefs.

These having been laid together, and noted in a book, were passed on to another set of men, who wrapped the whole in a sheet of "merikani." The parcel was then firmly bound with a rope till it became as hard as wood, though retaining some elasticity—the operation of roping being assisted by a thorough beating with a wooden club. The bale in this form was passed to a fourth party, who sewed it up in rough matting, to ward off rain and preserve the contents from injury by the rough usage it would be subjected to. The matting also makes stealing more difficult. I next painted its proper number on the bale, and the process was completed by Johnston writing down the number and contents of the bale in his note-book, so that when any particular article was required, all that we needed to do was to turn up the list of goods in our note-book, and find out the number of a bale that had the required article.

Each load, when finished, averaged about sixty pounds weight, and had the form of a cylinder, three-and-a-half feet long and one foot in diameter. This load each porter was supposed to carry on his head or shoulders from two to eight hours a day.

In addition, he had to carry his gun, mat, cooking-pot, or other personal property, which would commonly bring his actual load up to eighty pounds.

The beads were put into bags carefully marked; while the copper and brass wire was tied in coils at the extremities of a pole six feet long, and carried on the bare shoulder. Miscellaneous articles, stores, personal effects, &c., were mostly packed in boxes, which, from their unyielding character and angularities, are hated by the porters. Our stores for personal use consisted of a small supply of tinned meats and soups, sugar, coffee, and tea, and a few other European articles. From England we had brought a tiny medicine-box; a rifle, a gun, and a revolver each, and two small tents.

In the selection of our tents we soon found we had committed a grave blunder. In travelling in a country like Central Africa, where, during one half of the year, there are daily deluging rains, and during the other half a withering sun which strikes through a flimsy covering with burning effect, where, moreover, there are remarkably sudden changes of temperature (amounting sometimes to 30° within the hour, as I myself have noted), it is of vital importance to have a roomy tent. Fevers and other diseases are only too common; and to be shut up ill in your tent, with your face close to the roof and your hands touching both sides, means an addition to your tortures of no mean account.

Mr. Johnston had a tent which for size would

have been suitable enough for a short excursion, but was totally destructive of comfort when one or two years had to be spent in it. It was but six feet long, five feet broad, and less than six feet to the top of the sloping roof. Neither chair nor camp-table could find a place in it; and in rainy weather or during illness there was nothing for it but to lie down in bed and study the roof in a horribly close, stuffy atmosphere, that was almost choking.

This was bad enough; but my tent was ten times worse. It was an absurd little thing, into which I had to crawl on hands and knees; I could not even sit upright in it, and when lying in bed my face was almost touching the roof. The material, moreover, was so flimsy that I had not used the tent a week before it began to leak at every pore during rain. For six weeks I used that tent in all weathers, and if I had had to use it for another six, I would certainly have had the honour of becoming, not a martyr to geographical science, but the victim of a stupid blunder.

It seems to me that more travellers have died in East Africa from such blunders and avoidable causes, than from all other agencies put together. The troubles, failures, and deaths, for instance, which have so often marked the progress of the Belgian International Association's Expeditions, whose acquaintance we shall make further on, are easily traceable to the marked absence of all care and common sense, not only in the sending out of men, but in the fitting out and organizing of their caravans. The leaders have arrived at Zanzibar, either totally ignorant of, or supremely indifferent to the requirements of the work before them. They have hurriedly got together a nondescript collection of men and goods, and started for the interior as if everything lay smooth before them. The next we hear of them is that they are wandering like lost sheep in the wilderness, paralyzed by fevers, deserted by their men, in trouble with the natives, and plundered by their own porters. Loud complaints are then made about the frightful climate, the perverse savagery of the natives, and the untrustworthiness of their followers. Sufficiently convincing proofs of the folly of such proceedings will present themselves in the course of our narrative.

In the intervals of our bale-making, we took every opportunity of interviewing Arabs and natives who had been in the countries through which we proposed to pass. Little knowledge, however, was gained, as our route was altogether away from the main lines of trade, and what could be learned was not of the most encouraging nature.

When our preparations were well forward, it became evident that we were likely to have some idle time on our hands. The rainy season was still far from being past, and until it was over we could not proceed on our journey. How could we find profitable employment for our leisure? A happy thought struck Mr. Johnston. Why not have a trip to Usambara?

As Mr. Johnston's lieutenant, I hailed the proposal with delight. Such an excursion promised to be not only most pleasurable, but in our special circumstances most profitable. It would give us a real glimpse into the interior of the mysterious continent, and afford us some conception of our future field of work. It would give us also a gentle and yet wholesome introduction to the difficulties of managing a caravan—difficulties which we should soon have to encounter on a much more extensive scale.

The trip would occupy nearly three weeks, and forthwith we set about such preparations as were necessary. These of course were not formidable, and in our eagerness to get into action we very speedily had all in readiness. Chuma soon provided half a score of stalwart sons of Africa to carry our indispensables. These consisted for the most part of clothes, cooking utensils, articles of barter, and presents for the chiefs on our way.

On the 24th of February, we had everything put on board the dhow which was to convey us across to the mainland, and which we purposed to meet at the north end of the island on the following day. We bade a playfully impressive farewell to our friends in Zanzibar, and on the morning of the

25th, started on our walk to Kokotoni—a distance of about thirty miles. We did not forget how Speke had been twitted for travelling with a hundred musketeers over territory which a German missionary had safely traversed, "weaponed only with an umbrella," and so we sallied forth in no more warlike array than that peaceful preacher. The road to Kokotoni lay through cultivated ground, amid mango, clove, orange, and cocoanut plantations, which make up the main features of the vegetable scenery of Zanzibar. These trees, though they represent only a few species do not by any means convey the impression of sameness. On the contrary, there is a most pleasant variety of form, and colour, and grouping. Indeed the massive mangoes in combination with the cocoanut-trees are in themselves quite kaleidoscopic in the almost infinite diversity of charming scenes they produce.

We reached Kokotoni as the sun was setting, and found the dhow waiting for us. But before getting on board we could not resist the temptation of having a plunge in the sea; nor did we find that luxury the less refreshing from the exciting thought of possible sharks, which are abundant here. Having disposed of supper, which we found ready for us, we unrolled our mats, and speedily were enjoying a well-earned sleep on the half-deck of the dhow, as it glided out from the island. When we awoke in the morning we found ourselves near the mainland; but as the wind fell to a dead calm, which lasted nearly all day, it was almost night ere we reached Pangani. There we landed in the midst of a drenching rain.

The town of Pangani stands at the mouth of a river of the same name. Its position is most unhealthy—situated as it is on an ancient swampy delta of the river, very little raised above the level of the sea, and surrounded by high banks. It is almost hidden by plantations of cocoanut-trees from the leaves of which the natives construct their huts. The divisions of the leaves are platted so as to resemble basket-work, and thus a light airy dwelling is formed well suited for such a climate. Besides these native huts, there are a number of prison-like houses, built of coral rock and occupied by the Arabs. These are placed regardless of all order, forming irregular filthy lanes, where all sorts of garbage are thrown to lie and rot.

Having donned a dry suit of clothes, we proceeded to make the necessary ceremonial call upon the governor of the place. He received us seated in state in a dirty shed, surrounded by his kinsmen and retainers in all the glory of turbans and highly coloured garments. We presented the letter of introduction with which Dr. Kirk, our Consul-General at Zanzibar, had furnished us. We made the usual salaams; but what to do or say next was a problem we were quite at a loss to solve. For nearly an hour we sat in silence—silence which to us was at once amusing and embarrassing. It was with genuine pleasure we found our meditations at last broken by the appearance of the customary sherbet and ridiculous cups of coffee, which formed a welcome variety to our trying visit.

That night was spent in futile attempts to sleep amid myriads of mosquitoes, and (I may add) other unmentionable, but no less exasperating, insects which accompanied the pillows sent us by the governor. We hailed with joy the morning light. As the caravan was not to leave till the afternoon I foolishly determined to let off a little of my superfluous energy by a few hours' exploration of our surroundings. I first tried the overhanging forest, but only succeeded in getting my clothes torn, and my person painfully scratched. I then, in a second suit, tried the mangrove swamps, the haunt of the never-silent frog, and the feeding-ground of beautiful kingfishers and herons; but I soon emerged from them one of the most soaked and blotched of individuals, and with decidedly fresh views on the subject of African travelling. My education was rapidly progressing.

At three p.m. we elevated our conspicuous umbrellas and evacuated Pangani. Following a native path, we soon reached a carefully cultivated tract on the top of an ancient raised beach which skirts the east coast of Africa. Here the variety and beauty of the birds attracted our attention. There were jays, bee-eaters, weaver-birds, and sun-birds in great numbers; while a species of thrush displayed its musical powers in a fashion that surprised us, accustomed as we were to speak of the absence of singing birds in the tropics. Here, too, the ants exhibited working powers which were nothing less than marvellous. We speak of these tiny creatures as models of industry; but how little do most people appreciate their actual capacities! We passed mound after mound raised by these little builders, and so large that many an ancient Briton might have felt honoured to be buried beneath them. It was no uncommon thing, indeed, to see a good-sized tree growing on the top of an ant-mound.

We had not proceeded far on our way when rain began to pour; and when you consider that the native path is simply a narrow track through overhanging grass, generally taller than a man, you may be sure the change of weather neither increased our comfort nor rejoiced our heart. Still on we pressed, and in little more than two and a half hours we reached what appeared at a short distance to be an impenetrable wood. We found, however, that our path led us into a perfect tunnel. Hurrying through this, we were suddenly ushered into one of the most romantic little villages that can well be imagined.

Picture to yourselves a circular space two hundred feet in diameter. Round it plant the densest forest you can realize. From every available point in the circle of trees hang manifold festoons of creepers binding the branches into one unbroken mass of foliage. Fill the interstices between the trunks of the trees with greenest shrubbery. On the clear space build a number of mud huts, thatched with grass, and of all shapes—square, round, oval, composite—not one to be the same as another. Add some life to the scene by throwing in a few negroes, dogs, goats, and hens, and you have before you the little village of Madanga.

We occupied a native hut for the night. We of course tried to sleep, but with no better result than at Pangani. The silent hours were mainly spent in savagely wishing that we could see the unwelcome companions of our midnight unrest pinned in some insect collection. We captured a few, but of course they were too much crushed to be of any value to the most enthusiastic collector.

Resuming our march on the morning of the 28th, we were at the outset treated to an involuntary shower-bath as we pushed our way through the tall wet grass. The country at this point was but sparsely wooded. Still it presented many botanical novelties. There were the baobabs—thorough vegetable Bohemians in their eccentricities of form and their isolation from all arboreal society. Their appearance suggested enormous inverted turnips with the rootlets representing the branches. Among other vegetable curiosities were the hyphœne—the only branching member of the palm family; the fan-palm, with its oddly bulging trunk and its continuously flapping fan-shaped leaves, looking as if it were finding its place too hot for it; and the still more remarkable euphorbia-tree—a huge cactus-like plant with many branches, but not a leaf. Of this last monstrosity nature would seem to be ashamed, for she almost invariably imprisons it in some shady clump of trees where it is utilized by many charming creepers, which pleasingly cover up its unsightliness while exhibiting their own grace and beauty. Parasitic plants are exceedingly common. Prominent among these is a species of fig, which frequently grows so large as entirely to bury up the tree upon which it lives. Its method of working is as follows:—Whenever a tree gets wounded the parasite selfishly settles down like a vampire upon the place, and commences its ruthless work of destruction. Roots grow down, and gradually embracing the trunk become welded together. At the same time the parasite rapidly shoots up like an ordinary tree, till it raises its showy branches aloft with all the pride of a forest giant.

Apart, however, from these botanical curiosities there was much that was attractive to gladden the eye. I was greatly impressed by the wonderfully picturesque effects produced by the profusion of leafy creepers hanging on the trees. These frequently, in their more massive forms, reminded me of some old baronial ruin which the kindly hand of nature had, with sympathetic touch, covered with greenest ivy to preserve it from the further ravages of time.

Strange to say, throughout the whole of this day's march hardly a single bird was seen, and not one flower brightened the pathway. The latter fact astonished me very much; for do we not naturally connect the tropics with all that is rich and rare in floral beauty? Indeed, during the whole of our three weeks' trip I did not see a single flower worthy of a glowing sentence of description.

Our second day's march from Pangani ended at sunset at the village of Kwamakumba, which, like Madanga, is situated in the heart of an impenetrable forest. In this place clothes are manifestly at a discount, especially among the ladies, who have nothing but a small piece of cloth hanging from their waists. This they never wash; nor do they ever sew rents. Hence their "dress" is generally one of simple shreds and tatters, kept together no one knows how. They prefer, moreover, the very thinnest of materials. Even gauze is not considered too airy a fabric. One young damsel sported as the sole apology for a garment the remnant of an old fishing-net.

On the third day from Pangani we reached Magila, where the Universities' Mission has established a station. Here a number of missionary gentlemen make themselves martyrs to the conventionalities of their Church in a manner which greatly amused me. Perhaps their consistency may be worthy of admiration, but I confess it tickled my fancy to see these men, with the thermometer at 90° in the shade, wearing long black priestly garments hanging to their feet, ropes round their waists, and shovel hats, in which they pushed through forest and jungle, plunged through swamp and stream, handled the axe or the spade, and finally held divine service in their fancifully decorated chapels. I have no desire to convey the impression that this ritualistic mission is doing no good in Magila. Undoubtedly the missionaries are earnest and devoted men, whose soul is in their work; but in their methods there is unquestionably much that is absurd, if not worse. They certainly succeed in raising the vulgar admiration of the natives by their showy vestments, ceremonial processions, candles, and altars; and by substituting a cross when they take away the "dawas" or charms from the negroes they convey an idea of the Christian religion, which suits naturally superstitious minds. But whether all this is calculated to produce the highest results I very much doubt.

We found Magila situated in a charming and populous district at the base of the mountains of Usambara. As we were eager to get some glimpse of the country we had come to see, we resolved in the afternoon to ascend to the top of a prominent peak which rose behind the mission station. It was a somewhat hard climb up three thousand feet, but the labour did not lessen the enjoyment. The whole mountain was covered with rich forest, with the exception of a few cleared spaces where a number of natives had taken up their abode for greater security, and down its side dashed a succession of cascades half hidden by huge ferns and bananas. Imagine the pleasure of meeting among the former a number of old friends, notably the bracken.

While peering about like an adventurous, inquisitive monkey, I came suddenly upon a strange and grotesque procession. Dancing along in single file were a number of natives. Their faces were smeared with ashes, while their clothing consisted of grass kilts, with a thick band round their waists, from which hung various curious appendages to represent tails. This latter I have since found to be a common practice all over Inner Africa. I learned on inquiry that a number of youths had just assumed the tribal mark, and that this peculiar ceremony was a celebration of the event.

On our return I was somewhat surprised to hear certain shrill sounds, usually supposed to be heard principally in the Highland glens of Scotland. Anything so suggestive of the bagpipes naturally made me prick up my ears, and awoke native memories within me. However, I discovered that the instrument rather resembled a clarionet in shape, and was called by the natives "zomiri."

The following day being Sabbath, we remained at Magila, and had an opportunity of observing the mission services, both in English and in Swahili. The native converts sang the hymns very pleasingly; the discourses were simple and appropriate.

On Monday we were up with the first streaks of dawn. Having a very arduous task before us, we left everything but what was absolutely indispensable. The missionaries had given us a very glowing account of the sights that awaited us, and we felt in a very joyous frame of mind when we began our actual march into Usambara—away into the glorious Usambara mountains! We found that the country had not been described in exaggerated terms. In fact, it would be almost impossible to speak too enthusiastically about it. When I think of describing the sights of this day, I find that words fail me. Well-merited superlatives are at the tip of my pen; but of what use is it to besprinkle them over the page? They could convey to the reader no definite or adequate idea. They could only indicate how much I was impressed.

Our route skirted the south end of the Magila Mountain, and led us at first through an undulating country, covered with a dense wood of short trees, shrubs, and creepers. As we moved silently onward in single file, we observed that the country was becoming more and more uneven, and cut up into sharp hills and deep dales. The trees also became more and more lofty, until I felt I was fast approaching that ideal of my dreams, a real primeval forest. What a walk we had of it! We would mount up, and seemingly ever upwards, gasping for breath, and yet eagerly peering on this side and on that, in a perfect ecstasy. Then we would descend as far into deep, bowl-shaped depressions, and work our way through narrow gorges until a feeling of awe would steal over us as the sun disappeared from view, and silence and gloom reigned supreme.

It was, indeed, a marvellous forest; every tree a veritable giant, rising with bare trunk, as if struggling for the free air of heaven, to a height of from seventy to a hundred feet before branching, and then forming a parachute-shaped crown, through which the rays of the sun in vain attempted to penetrate. Little less gigantic than the trees were the inevitable creepers. None of your slender convolvuluses, or passion-flowers, or ivies; but massive fellows, thick as a man's thigh, and two or three hundred feet in length, hanging aloft from tree to tree, or wriggling and twisting up their stems. Everything was strange, and grand, and colossal! In such an amazing development of vegetable life, one looked in vain for something of simple beauty—some flower, some pretty bird or insect. In the damp, dark recesses of that mighty forest, what of light-loving beauty could be expected to exist? Nature, too, as if exhausted in the rearing of these arboreal giants, presented nothing animate to enliven the gloom. Not a sign of animal life was visible. And, but for the distant hoarse cry of a solitary hornbill, or the occasional croak of a frog, one might have supposed none existed.

Through these sombre scenes I was wonderingly

winding my way, considerably in advance of the rest of my party, when all at once I became aware of a mighty humming sound, apparently proceeding from some distance ahead. It increased as I advanced. At last I emerged from the forest, and had my curiosity satisfied. There, in an open, circular piece of ground, a crowd of women were holding a market. What other assemblage of animated beings could have produced such a noise? My unexpected appearance caused universal silence. Fear was depicted on every face. Yet, like suddenly startled deer, they must needs pause a moment and stare at me—though I was evidently supposed to be a ghost. At last I observed preparations for a grand stampede, whereupon I stepped forward with all the bland style of a member of the Peace Society, and elevated my umbrella to show that it was not a more murderous instrument. The effect was surprising. I evidently won the hearts of the fair sex at once. Unfortunately this confidence did not last long. I sat down to take notes, and the appearance of the book made them look once more suspicious. But when I produced my watch to mark the time, they broke instantaneously and cleared out, in spite of every protestation I could make.

Some distance farther on I espied a village perched in a position that might rival the Castle of Edinburgh. As no hard piece of work daunted me in those days, I clambered up. Finding a rather frightened-looking group at the top, I tried my late successful ruse, but with a very different effect. On elevating my umbrella they precipitately took to their heels as if I had fired a cannon at them. Following them up, I found them ensconced behind rocks armed with bows and arrows, which they seemed not at all disinclined to use. As they were evidently not disposed to receive a pioneer of civilization within their gates, it was worse than useless to attempt a forced acquaintance. So I voted them barbarians, and indignantly, though with some trepidation, retired.

At mid-day we rested on the banks of a large stream, called the Zigi. In its waters we enjoyed a delicious bath after our hard morning's march, in which we were impressed with the truth of the lines—

> "Those high wild hills and rough uneven ways
> Draw out the miles!"

Resuming our journey, we kept steadily at it till, near the close of day, we reached the top of a mountain, where we all threw ourselves down on the ground, dead beat. Here we were within sight of our destination for the day, namely, the village of Hemasasa. But, before entering it, our guide satisfied his sense of the importance of the visit by giving a salvo of two shots from his old Tower musket, which, charged as it was, made him spin round in rather an alarming fashion. Having thus frightened the people of the village out of their wits, he allowed us to proceed.

Hemasasa proved to be a delightful place. Commanding a glorious view of mountain and forest, and situated as it is at a height of 3000 feet, it well merits the application of the lines—

> "Fair is that land as evening skies,
> And cool, though in the depths it lies
> Of burning Africa."

The sights and scenes of the following day's march may be fitly described in the words of an American humorist as being "like the preceding, only more so." We had actually to wait an hour after the sun rose before we had sufficient light to enable us to proceed, as our way led through an exceedingly deep and narrow gorge. Guided by the chief's henchman, we at last set out. Descending a slippery and almost perpendicular path, we reached with difficulty the bottom of the gorge, a thousand feet beneath the village. Through this we groped our way in a light sombre and gloomy.

The grandeur and peculiar forms of the forest trees continually evoked new admiration, rising as they did from 100 to 200 feet in height, with trunks of proportional thickness. Some had the appearance of being buttressed with huge slabs, eight to twelve inches thick, and extending up the trunk ten feet, sticking out all around like rays. When carefully cut off, these buttresses form ready-made planks for the natives. Another species of these monster trees seemed as though it were set upon a pedestal, so suddenly does it become contracted in girth about ten feet from the ground. Many of the trees had a crown of leaves, each one of which was at least three feet long and a foot in breadth.

But all these arboreal wonders were instantly forgotten and neglected as my eye lighted on a lovely group of tree-ferns growing beside a rocky stream, with straight stems twenty feet high, topped by delicate soft green crowns of fronds. The sight was in every respect most charming and memorable,—the dashing waters flowing through the deep picturesque valley, and that exquisite cluster of ferns, sheltered alike from rude blasts and burning sun by the great guardian forest, which cast a dim twilight shade around. This part of the country proved to be a very paradise of ferns. On all sides the ground was lighted up by them. They peeped from beneath each rock, they beautified its every crevice, and tenderly toned down its rugged angularities. They clung to the stems of trees and hung about the branches. Such a lavish, natural adornment is, indeed, rarely to be seen.

It was quite with a feeling of regret that we found ourselves about mid-day at the terminus of our trip. This was Hendei, the capital of the surrounding country, and where Kibanda, the chief, resides. As at Hemasasa, our guide had to announce our arrival by the usual salvo. Much to our alarm we observed him cramming the barrel of his gun with powder, and, to confess the truth, we took refuge behind trees, in case of an explosion. This ceremony safely over, we marched into the village, and were conducted to a fine large hut set apart for distinguished strangers. Shot after shot saluted us on every side, and often in such dangerous proximity that it required some nerve to preserve a dignified demeanour. However, we succeeded admirably in standing fire, and at last got settled down in our hut, surrounded by a staring crowd.

After a refreshing wash and some dinner we were ready to receive Kibandi in proper form. In a short time he arrived, dressed in Arab fashion, evidently much to his own discomfort. He had rather good features, and was not very dark in colour. Mr. Johnston laid a tempting array of presents before him. He seemed to be thoroughly pleased, and in the afternoon he sent us a present of a bullock and other food.

As soon as this interview was over I set off with two guides to the top of a high mountain near the village to collect plants. The ascent required a vigorous effort, but was soon accomplished. While collecting the plants I was the subject of innumerable jokes. My guides could not understand all this work, and made themselves very merry at my expense. However, as I did not understand them, I let them laugh.

At night there was a dance in honour of our arrival. The boys and girls led off with some pleasing movements. The men followed with a fiendish "breakdown." They seemed to go almost mad as the drums thundered out their infernal din in a way which would have accorded well with the performances of "Auld Clootie" in Alloway Kirk, and, as on that memorable and authentic occasion in the history of Tam o' Shanter, there was certainly no lack of "cutty sarks" or "eldrich screeches."

In the middle of the night after our arrival at Hendei I was wakened up by a very peculiar sensation. I felt a creeping all over me. I thought at first that I was dreaming; but hearing at last various expressive interjections in English, and then in Swahili and Arabic, followed by a rush for the open air, I became alive to the situation. I found myself literally covered with ants. I ran frantically out of the hut, feeling as if a million needles were piercing me in every part. I danced and writhed about, as did Johnston and our men. I tore frantically at my hair, and reduced my clothes in a twinkling to a minimum. The exasperating little creatures swarmed in perfect myriads over us—beneath our clothes and into our hair. It was only after a two hours' battle we contrived to get clear of them. We felt as if we had passed through a "baptism of fire."

Next day, while Mr. Johnston ascended Hendei Mountain to make observations, I sallied forth to see the country and shoot birds. We had a somewhat trying time of it. To any one who has been in a tropical forest I only require to say that I went off the pathway, to describe my difficulties. Any attempt to move in such a place without a beaten track means a prolonged and fruitless struggle with tall grass, matted bushes, and interlacing creepers. Your clothes get literally drenched with streams of perspiration, and you may be thankful if you do not get them torn off you altogether. Still, one has always a certain pleasure in facing difficulties, and quite an exhilarating feeling usually took possession of me when I got into such places. In this instance I got few specimens in return for all my labour, and fewer blessings from my guides, who, being accustomed to an easy do-nothing sort of existence, could not understand the philosophy of such a squandering of energy.

In the evening I was again on my feet, this time for a monkey-hunt, as I had not yet seen one in its natural habitat. When we set out the sun was sinking beneath the horizon, adding to the natural gloom of the forest. We proceeded rapidly for some time till we—that is to say, myself and two guides—came to a promising place. Every sound was hushed, and we peered about like villains with bated breath for the objects of our search. The stillness was deathlike, and as the darkness deepened an "eerie" feeling crept over me. Yet I felt a strange fascination in the scene, as I dimly discerned the two naked savages gliding about like evil spirits among the trees, without the rustle of a leaf or the noise of a footfall. At last they warned me by signs that game was in sight; but I had only time to get a glimpse before it disappeared. As the light faded away, and we still glided about in unbroken silence, I felt more and more awed, till I quite shivered. We were on the point of beating a retreat, when suddenly an object was observed to rustle among the leaves of a tree. I aimed and fired, breaking the awful quiet in an alarming manner. As the report of the gun rolled away through the forest vistas, hoarse cries from hornbills and terrified screams from unseen monkeys rent the startled air, while with crash after crash a large baboon, shot through the breast, fell from branch to branch, till it struck the ground with a dull thud, dead, its eyes fixed, teeth set, and mouth foaming. I felt considerable compunction at the deed I had done.

By this time it had become so dark that we could hardly find our way back. One of the guides ran on before to announce our approach, and to sprinkle ashes on the gateway to ward off any evil. We were received by the villagers like heroes returning from a victorious campaign. My hunting exploits were doubtless the subject of many extraordinary stories round the camp-fire that night.

We had now accomplished the objects of our excursion, and seen the wonders of Usambara. So, early in the following morning, we began our return journey. We said our farewell to Kibanda, but begged to be excused the trying effort to bid adieu to his fifty wives. Taking a different route from that by which we had come, we pushed rapidly through forest shades, and over many a rugged hill and picturesque dale, finally camping for the night at a small village a short distance from Magila.

We slept in the open air, and, not having the poetry worn out of us at this early stage of our career, we lay long awake, watching the clear starry skies, and the great sighing forest, lighted up, as if with fairy hand, by the flash of those earth-stars, the little fire-flies; while the cicadæ held high revel, filling the air now with the tones of tiny silver bells, and now with clear flute-like notes—a perfect medley of exquisite sounds—contrasting with the weird hooting of the owls and the sage voice of the frog. The deep though distant roar of the lion made the effect all the greater, as we pictured him at his feast. At last the stars began to take erratic courses, and everything slipped into dreamland.

It was not long before we were rudely recalled from the realms of romance. Rain came on and compelled us to creep under our waterproof sheets, which saved us from the falling moisture, but drenched us with perspiration. I was just on the point of sleeping for the second time, when an ear-piercing yell rang out through the air, and made me spring to my feet and instinctively seize my umbrella, as drowning men clutch at straws. A leopard had carried off a dog from within a few feet of where I lay!

Three more marches brought us to Pangani,

where I made the first propitiatory sacrifice of my health to satisfy the evil genius of Africa. In other words, I became ill with fever.

Early on the morning of the 13th of March we got on board our dhow, and with a fair breeze soon ran across to the Island of Zanzibar. There, however, we got the wind right in our teeth. Finding it impossible to reach the harbour, we landed in a canoe under rather dangerous circumstances. On shore I found myself so ill that it was with the utmost difficulty I entered the town. Under the Consulate doctor, however, I quickly recovered, when Mr. Johnston also became ill, and suffered severely for some time.

These troubles by no means alarmed or discouraged us. They were not unexpected, and they formed a seasonable discipline. We had thus, then, in a sense, completed our apprenticeship as African travellers. We had gained a fair notion both of the trials and labours we should have to encounter; and our experience only made us more confident and enthusiastic than ever in proceeding with our greater undertaking.

This preliminary trip was of undoubted value to us in many ways. It corrected our mistaken notions concerning African travelling. It enabled us in some degree to measure our individual capabilities for the work we had undertaken. It helped us to draw a pretty accurate line between the possible and the impossible in exploration, and thus was the means indirectly of saving much precious time.

Unfortunately Johnston's seasoning did not end with his fever. The process was further prolonged by an abscess in his ear, which gave him much trouble for several weeks.

Much has been written for and against the belief in the reality of a seasoning fever. I myself am convinced that there is no such thing, in the sense that you are rendered not liable to take another fever after having gone through the first. But of this I am certain, that the new comer who takes fever soon after his arrival is the person who is most likely to stand the climate. A rapidly contracted fever is generally not a serious one, and succeeding fevers become less dangerous and virulent. On the other hand, a person who comes from England to such a place as East Africa, and does not take fever for a long time, and seems to find the climate agree with him, is almost certain to have the fever in a dangerous form when it does come. Though the person may not have noticed the fever poison, it has been slowly and insidiously working itself into his system, till it has fairly marked him for its victim.

After my recovery from the fever I was plunged into an exceedingly lively state of mind and body by what is known as "prickly heat." What a glorious scourge this trouble would be to any sinner anxious to do penance for the mortification of his flesh! Under its influence one feels as if he had his coat full of needles, with the points all directed inward, so that if he moves, or even sighs, they are ready to pierce him with a million stinging wounds. If he rises from a chair, or bows politely to a lady, he can scarcely refrain from a variety of unseemly grimaces and expressive interjections. He cannot even perspire without imagining himself sweating blood. The social cup of tea as it is slowly imbibed seems to ooze out of every pore in the shape of needles. Laughter itself is turned into pitiful contortions, and even the lover dares not indulge in a sigh "o'er his mistress's eyebrows." The annoying ailment, while it lasts, seems, indeed, to rob its victim of every pleasure, and to make all life a bore, so that he is glad to escape into solitude. Such was my experience for nearly a fortnight. Fortunately, however, it very seldom troubled me afterwards.

By the time our health was quite restored the worst of the rainy season was over, and we began to look forward with impatience to our start.

To make sure of a successful commencement, Johnston, who in his eager enthusiasm left no stone unturned, determined to visit Dar-es-Salaam, the town on the mainland from which our explorations would commence, in order to ascertain by personal inquiry and observation what would be the best route, as there appeared to be more than one. On landing at Dar-es-Salaam, with Chuma and a few men, he met Mr. Beardall, a remarkably energetic young surveyor, employed by Sir Fowell Buxton and other British philanthropists to construct a road in the direction of Lake Nyassa, for the purpose of introducing to the natives the benefits of trade and civilization. Mr. Beardall, by his knowledge of the surrounding country and of the natives, gave Mr. Johnston much valuable assistance in his researches.

They both proceeded up the road as far as it had been made, finding it at that time of the year submerged almost along its whole length by the rain.

From the natives of the country Mr. Johnston received much information regarding the best routes, and also regarding the various tribes through whose territory the Expedition would proceed. As the result of his survey and investigations, he decided we must strike more to the south from Dar-es-Salaam, in place of going along the road, where the food was scarce, and where the water would seriously impede our progress. The start of the Expedition was also delayed for a fortnight, as the country was still much flooded and the crops scarcely ripe.

The date was at last fixed, and we began to hurry our final arrangements. The porters were all taken before Dr. Kirk. After having the agreement read over and explained, accompanied by a judicious hint as to the penalties following desertion or stealing, they were formally engaged. Each man received four months' wages in advance, the rate of pay being five dollars a month, with rations on the march.

The porters were evidently much impressed with the part which Dr. Kirk took in our arrangements. He is looked up to with almost superstitious reverence. Hence the interest he manifested in the Expedition gave it a coherence and a character otherwise unattainable. They felt it was quite a British Government affair, and held up their heads accordingly.

On the 13th of May we went to pay a farewell visit to the Sultan, and to thank him for his kindness in having lent us one of his steamers to convey our goods and men to the mainland. This visit being more of a private character than on the former occasion, we were not received with the same imposing display of soldiers. In all other respects, however, it was similar. A map of Africa was brought out, on which Dr. Kirk indicated our proposed route. In bidding us farewell the Sultan graciously expressed the hope that we should have a good journey, discover all that we went to seek, and return in safety. He then shook hands with us, and we left, delighted with the real kindness and pleasant courtesy of his Highness.

On the following day the Old Consulate presented a picturesque and animated scene. Thither at early morning the porters flocked in their best clothes, and accompanied by their wives carrying the mats, pots, &c., which were to be used on their journey. As each arrived he was passed into the court of the house, and was on no account allowed to go out again, a special guard of men being stationed at the door. After breakfast all the bales, boxes, and other baggage were carried to the boats, and sent on board the steamer "Star." This operation occupied us till midday. It was then the turn of the porters to be shipped. Hitherto they had been in a very jolly and noisy condition, though among the women a quieter and more subdued tone prevailed. The latter were practically being deserted by their husbands, and would have to find their living during their absence as best they might. They had thus good reason for looking anxious and sad.

Johnston stationed himself at the door, and a double line of picked men was formed to the water's edge. As each man's name was called he was marched down, accompanied by his wife and "traps," to the boats. There he parted with the former in a somewhat careless and unfeeling manner, she falling back among the crowd, not without signs of strong emotion. Some of the poor creatures, indeed, broke down entirely, and little wonder. After the terrible loss of life in connexion with Stanley's expedition and others, it seemed to be something of a forlorn hope to join a European caravan and travel through unknown lands.

Much to our satisfaction the men were all passed on board before sundown without one missing.

It may sound rather strange to British ears to say that many of these porters were Zanzibar slaves.

The matter is very easily explained. Though we had slaves in our company, we did not engage them *as* slaves. They came to us offering to enter into engagements just like free men, and they were accepted and treated as though they were free men. What agreement they had previously entered into with their owners we knew not, and did not trouble ourselves to inquire. With that we had nothing to do. We simply accepted their offered services, and hired them as men fit for our purpose. Of course as slaves their owners had a right to restrain them from leaving Zanzibar without permission, and in one or two cases this right was exercised. It would be next to impossible to raise a caravan without in this way having slaves included in it.

The 15th of May opened with one of those cool breezes so characteristic of Zanzibar. We had passed the night at the Old Consulate, and we were up early, with nerves highly strung, as we felt we were at last on the point of setting to the great work before us. We crossed over to the Consulate for an early cup of tea, which we drank in silence. We felt almost as if we were leaving a pleasant home of our own when we shook hands with Mrs. Kirk and the children, with whom we had romped about. We at once went on board, accompanied by Dr. Kirk. The anchor was weighed. The men gave a farewell cheer to their friends on the beach, and we slowly steamed out of the harbour. H.M.S. "London" dipped her flag in our honour as we passed, and many were the waves, and answering waves, of handkerchiefs as we rounded Shangani Point and left Zanzibar behind us.

The all-absorbing question was, What will our fate be in the Dark Continent?

CHAPTER III.

DAR-ES-SALAAM TO MKAMBA.

As our good little steamer, the "Star," kept on its way over the waters of the strait which separates Zanzibar from the mainland, the ever-green island of the Indian Ocean, where we had spent so many pleasant days, began to disappear as in some great transformation scene. In the fast deepening haze of distance each well-known landmark became indistinct, till at last nothing was to be seen but a few isolated cocoanuts peeping above the watery horizon.

But, as the island receded into dimness, the mainland rose before us with increasing distinctness, and gradually assumed shape and form. From the shore stretched a monotonous, green, unbroken landscape; and away beyond, the bold and picturesque mountains of Duthumi and Usagara rose abruptly from the low-lying malarious plains, promising pure air and cool refreshing breezes.

As we near the land the ordinary scenery of the east coast of Africa gradually developes itself. There are the sea-loving cocoanuts—tall and ungraceful when old and isolated, but of fern-like beauty when young, with their waving crown of leaves set on a stem of appropriate length. The light green colour and feathery slender proportions of the palms contrast pleasingly with the massive forms and dark emerald tint of the mangoes. Beyond the belt of cocoanuts and mangoes there appears an unbroken mass of vegetation, in which you fail to discern either variety of colour, or difference in the component elements. It is simply a deary, monotonous prospect of leaf and branch.

As we gradually approach our destination, Dar-es-Salaam, we naturally look for some signs of a harbour. We see nothing, however, but an apparently unbroken coast-line, faced by dangerous reefs of coral, barely hidden beneath the water, and revealed only by the roar of the waves. The steamer is steered towards the shore by a narrow, zigzag, dangerous channel. A foaming sea breaks threateningly around us. So narrow is the waterway that a false turn of the wheel or a mistaken order would bring us at once to grief. The "Star" is commanded by Arabs, and the crew are negroes, so we naturally have our doubts about their skill, and watch the working of the steamer with breathless interest, as we hang over the bows and look out for breakers ahead. Nearer and nearer the steamer approaches the shore, and still we can discern no safe anchorage. At last, however, a slight opening like the mouth of a small river is seen. To this we steer, and slowly steaming into it a rapid turn is made, and fifty yards further on there lies the charming harbour of Dar-es-Salaam, like a placid lake, apparently cut off entirely from the sea, the vicinity of which is only evidenced by the constant roar of the breakers, beyond a narrow tree-clad strip of land.

Not far from the entrance stands the so-called town of Dar-es-Salaam, and in front of it we anchored, after awakening the echoes and apprising the neighbouring inhabitants of our approach by a shot from one of the ship's guns. From all quarters a motley crowd came rushing down to the anchorage in all the stages of dress, undress, and no-dress—Arabs in all the ponderous glories of turbans, kamerbands, and gorgeous clothes, Indians in their awkward though airy dresses, Waswahili in their kanzus, resembling night-shirts, and Wazaramo and other Washenzi (wild men) in simple loin cloths—all hastening to see the Great Man, Dr. Kirk, and his friends, and to present their salaams and greetings. In acknowledgment of our arrival the nondescript soldiers called Beluchees fired off the customary amount of powder from their old flint locks, that being the recognized mode of celebrating every interesting event, whether it be the birth of a child, a marriage, or a visit from a Mzungu. As we stepped on shore the acting governor—known as the Akhida—had the distinguished honour of shaking hands with us, and, as is usual, laying himself and all his possessions at our command. Let us but speak, and whatever might be our pleasure it should be done! On all sides we were pleasantly greeted with "Yambo bwana?" (How do you do, sir, or master?) to which we gravely replied, "Yambo sana" (I am very well).

Mr. Beardall, the Dar-es-Salaam road engineer, was also there to meet us. He led the way to a huge barn-like building which formed his headquarters and store-house. Here we found commodious and roomy lodgings during our stay.

Till within the last few years the harbour of Dar-es-Salaam was quite unknown. The whole coast had been surveyed by the Navy officers, and ships had passed and repassed without once suspecting that so complete and well-sheltered a harbour was in their immediate vicinity. The dangerous coral reefs and shoals which skirt the coast made mariners wary of approaching near the shore.

Syed Majid, the predecessor of the present Sultan, was the first to recognize its advantages. Struck by the fine shelter it afforded for shipping, and by the healthy character of the country, he determined to erect a residence there, and divert the trade of the interior to it. For this end he commenced building a town on the most liberal of plans, with straight and broad streets, well-constructed houses, and indeed everything on the most approved and enlightened principles. A house for himself and court, and a guest-house, were commenced and got well on for completion. Caravans were ordered to make Dar-es-Salaam their trading-port instead of Bagamoyo or Saadani, which were at that time the principal centres of commercial activity on the coast. Unusual bustle reigned throughout the place, and the rapidly rising town promised to be the finest on the east coast of Africa. Magnificent plantations of cocoanuts were formed, and natives and traders began to flock to it from all parts.

This revolution, however, was of short duration. Suddenly the news arrived that the Sultan was dead, and with this event a blight seemed to fall on the place. No Arab ever completes the unfinished work of his fathers. Such a proceeding, they imagine, would only bring disaster and ruin upon themselves; hence the number of half-finished tenements rapidly going to ruin which are to be seen in Zanzibar and elsewhere. Syed Majid's death, then, was fatal to the fast-rising importance of Dar-es-Salaam. Syed Bargash, who reigned in his stead, did not like the place, and withdrew all support from his predecessor's scheme.

The builders left their partially completed houses, the woodwork to be devoured by ants, and the walls to fall into unsightly ruins. Traders disappeared from it as from an infected spot. The caravans resumed their old routes to Bagamoyo, and desolation reigned supreme. Grass now grows in the streets. The tenantless houses give shelter only to bats, owls, lizards and snakes, and the whole place wears the aspect of an old battered and deserted city, instead of a new one just springing into life.

It will probably now be many years before the prevailing lethargy is dispelled, and a new reign of activity commenced. But certainly, from the advantages which Dar-es-Salaam possesses above most places on the coast, there can be no doubt that it will, sooner or later, become all that its founder wished it to be. Its chief drawback lies in the narrow and somewhat dangerous entrance to the harbour, which would make it extremely difficult of approach to sailing-vessels, though steamers, with proper care, have nothing to fear.

The country round Dar-es-Salaam has much the appearance of a great plain stretching away inland for thirty or forty miles, with the low Marui hills rising from it to a height of about 300 feet. This plain has originated from an upheaval of the bed of the ocean, which has taken place in comparatively recent times (geologically speaking), producing what is known as raised beaches. The sheltered creek forming the harbour has evidently been originally one of those very deep channels so common among coral reefs; and similarly many hollows which formerly existed in the coral have now become beautiful ponds on the plain, succeeding each other from the coast inland.

The raised beaches are two if not three in number. They rise like so many steps, though varying very much in breadth—here contracting, there expanding. They consist of coral rock and consolidated fragments, overlaid by brick-red sands, and clays, derived from the metamorphic rocks of the coast mountain-range, which, being rich in iron ingredients, give the *detritus* its peculiar colour. The lowest of the beaches rises from twenty feet along the sea-line to fifty or sixty feet further inland, there being a general slope upwards, as we see in any ordinary existing beach. Except for its vegetable covering, it is exactly as it was when at the bottom of the sea.

To the mind filled, as mine was, with the popular ideas current regarding tropical countries, their rich luxuriance, magnificent flowers, and marvels of the animal world, it gives an unpleasant shock to be planted for the first time in such a place as Dar-es-Salaam. The disappointment is intense when we find how far our ideal visions appear to have exceeded the reality. It is like a rude awakening from some pleasant dream. The mighty forests which one had anticipated, with their gigantic and strange trees, and their wonderful creepers and parasites, are non-existent. A clump of cocoanuts here, and a dense bush there, alone meet the eye. No gorgeous flowers, nor graceful ferns abound. We look in vain for the expected abundance of beautiful butterflies and strange insects. Birds seem to be rare, and few command attention. No monkeys gambol among the trees; no snakes hang from the branches, or noiselessly glide through the grass. Nothing seems realized but the existence of a hot steaming atmosphere.

These were my conclusions during the first few days succeeding my arrival on the coast. Everything fell far below my expectations, and my dissatisfaction was proportionately great; and I may here add that during the course of our travels I never saw anything to equal the glorious vegetation which we had found covering the mountains and filling the gloomy valleys of Usambara. However, when my first disappointment was over, and when I began to judge my surroundings by a more reasonable standard, my appreciation of them began to rise. I accompanied Dr. Kirk in one or two forest rambles, and found that even in the neighbourhood of Dar-es-Salaam there was more of interest and beauty than in my haste I had given it credit for.

In one of these excursions after birds we crossed the creek, near the surface of which black and white kingfishers flew with rapid regular beats. As we landed, a heron and a sacred ibis rose from the shore, and flew away to seek some more undisturbed feeding-ground. We tramped about for hours, occasionally noticing a modest pretty flower, or graceful creeper. Pushing through bush and grass jungle we ever and anon disturbed some butterfly or beetle, which fluttered or buzzed away, as we eagerly noted its appearance. From the swampy hollows arose numerous dragon-flies with their gossamer irridescent wings and fairy flight. Sometimes a centipede or a scorpion, and frequently huge millepedes, lay on the pathway. Few land-shells met the eye, though those that did occur made up in size for what they lacked in number and variety, the most prominent being the large shell of the genus achatina, often five inches in length, with dark stripes on a yellowish ground.

The birds, though far from abundant, were the most conspicuous class of objects. The most commonly seen were the weaver-birds. A black and crimson species builds its nest on the tall grass, where it generally sits, rocked to and fro by the breeze, and from which it rarely flies far. Its near relative, the yellow weaver-bird, is of a more lively and active disposition, preferring to weave its ingenious nest at the extremity of a cocoanut leaf, where, secure from the ravages of monkeys, it flutters and chatters like a busy housewife the livelong day. The pretty little whydah-bird was also not uncommonly seen flitting about, with its long tail-feathers undulating in the wind, as if assisting locomotion. The corythaix, with its crested head and brilliant colours, flew from tree to tree, or ran along the branches and watched intruders with a most comical look of intense curiosity. Frequently the tepe-tepe, a species of cuckoo, was heard tooting out a most exquisite liquid note from the low bushes which it frequents. Among other of the more prominent of the feathered creation there were orioles, pheasant-cuckoos, hornbills, fly-catchers, puff-birds, thrushes, bee-eaters, &c. Of all these there is none that deserves the name of songster, with the exception of the golden-vented thrush, which, more golden in song than colour, everywhere salutes the traveller with its joyous warbling. It stands out pre-eminently as the best feathered vocalist of East Africa, and completely refutes the notion that there are no good singing-birds in the tropics. In this enumeration of the birds most commonly seen in our first ramble about Dar-es-Salaam we must not neglect the African representatives of the humming-birds, namely the sun-birds. These exquisite little creatures are frequently found flitting about in search of insects, They curiously enough betray a great liking for stimulating juices like the "tembo," drawn as a spirit from the flower-stalks of the cocoanut-trees. They frequently become intoxicated by indulging too freely in this juice.

In this ramble Dr. Kirk secured a number of bird specimens, and also some young plants of the graceful casuarina, a feathery tree not unlike one of our firs.

Though our ideal of the tropics had given way to less brilliant views with regard to its general characteristics, yet in one particular our expectations were more than realized. Poets may linger lovingly over the summer's twilight and the "dusky sandalled eve" of temperate climes, and sing in raptures over its slowly unfolding charms and changing tints, but, after all, commend me to the "gloamin" of the tropics, with its incomparable, clear, ambient atmosphere, its exquisite loveliness and its refreshing coolness, enhanced by the burning heat and blinding glare which characterized the day. The tropical twilight is indeed short, but the charm is the greater. The prolonged pleasure of the temperate twilight is here concentrated and intensified, so that every sense takes in delight. It is then that the liquid notes of the tepe-tepe well forth from the surrounding bush, the wise voices of the owl and the frog break on the ear, and the cicadæ tune their fairy pipes for the nocturnal concert, with the sober glowworm or the comet-like firefly to lighten up the scene.

With such a twilight ended our first ramble. As the stars shone forth, and the evening breeze commenced, the doors were closed, the lamps lit, and social evening intercourse began.

During the night Mr. Johnston, somewhat to our dismay, became ill with fever, which seemed a rather ominous beginning. It was exceedingly fortunate that Dr. Kirk had accompanied us, as the invalid was thus able to have the best advice.

The night air of the mainland proved to be exceedingly delicious and cool after the continuous oven heat of Zanzibar. The temperature fell to 70° F.—a point it had never reached during all the time of our residence at the latter place. I thus enjoyed an exceedingly refreshing sleep. On the following morning Johnston was still worse, and indefinite delay seemed in store for us. Towards evening, however, he began to improve, and, much to our delight, he had almost recovered on the succeeding day.

The men and goods having all been landed safely, with neither the one nor other missing, we proceeded on the third day after our arrival to distribute the guns to the best of the porters. With these we were pretty well supplied. Johnston and I had each a first-class express rifle, a shot-gun, and a small revolver, and Government had supplied, for the use of the Expedition, thirty Snider carbines with the proper accoutrements. To these we added twenty old Enfield rifles, purchased at Zanzibar, and twenty-two of the porters mustered a heterogeneous collection of flint locks and other ancient weapons. In all we had seventy guns, a number which would have been formidable in any part of Africa, if the men had known how to use them, or had had sufficient courage to stand and aim when danger threatened.

The distribution of the loads was not such a simple affair. Every man struggled to get the lightest and easiest. None were satisfied with anything allotted to them, and all had endless excuses to make, so that an indescribable scene of noise and confusion arose. Each one tried to avoid the boxes, as being the most unpleasant to carry, from their want of elasticity, and tried to get a bale in preference. If we had been left to ourselves we would have been rather in a dilemma, but here Chuma was in his element. He danced about with indignation, seizing this man by the ear and that by the throat, and dragging him to his appointed load, while he volleyed out his threats, or lashed them with his satire. He was ably seconded by the headmen, who, thoroughly enjoying the pleasures of command, seemed to glory in laying violent hands upon mutinous porters. In an hour, however, the noise and confusion died away. Each man knew his load, and had apparently become reconciled to it, however much it might be against the grain.

And now, a word about the *personnel* of the caravan. First, of course, came Chuma. Of him we have already spoken. He acted as the manager of the entire caravan. As assistants, he had five headmen, of whom the most prominent was Makatubu. He acted as keeper of the stores, and was supposed to know all about them and where everything was to be got. He measured out the cloth or other articles of barter. A capital fellow was Makatubu, full of life and energy, and seemingly ever on springs, never grumbling at hard work. He had but one great defect. He had not the slightest influence with the men, and was utterly useless as a headman. In other respects he was almost equal to Chuma.

Next in order came Nasibu, tall and Arab-like, with many good points, yet who so assiduously acted as a mere hanger-on of the caravan as to be of little or no use beyond marching with Chuma to keep him company. His chief characteristics were his powers of sleeping and unlimited drinking, though he showed immense strength when thoroughly roused, and had always a good word of advice when a council was held.

Asikari, the third in order, had not seen so much of camp-life. He had been brought up about the Sultan's court as his Highness's slave. As became one of such rearing, he proved to be the dandy of the party. He had a genius for picturesque and effective dressing, which could not but evoke our admiration. He played the part of ornamental foreman to the caravan, dressed in a snow-white garment and a gorgeous turban, with a huge Zulu like shield in one hand and a spear in the other.

Next followed the two working officers—the giant-like Beduè, the biggest man in the party, and, strange to say, the best needleman; and Stamboul, formerly slave-driver, now adjutant and drill-sergeant in our caravan.

A more intelligent, faithful, and honest set of headmen could not have been collected in Zanzibar.

Next in point of rank we must put Litali the cook. This important personage was the most aggravating individual in the whole caravan, and certainly the most unmanageable. He had been brought up as a Christian at the Universities' Mission, and had there acquired a considerable stock of pious words and phrases, and learnt English and impudence to no small amount. He had a perfectly devilish temper, when roused, and it was long before I succeeded in "reorganizing" him. He, however, was not altogether given to the devil, and was specially valuable in jealously keeping all but himself from sharing in my little luxuries, such as tea and sugar. On the whole, his good points balanced his bad ones.

Of non-commissioned officers we had ten, known technically as Kiringosis, or guides. These are the picked men of the caravan, who always march in front, a distinction they jealously stick to. They are supposed to be the boldest and the best fighting-men, and as appearance is a very important element in African warfare, they have the coveted privilege of decking themselves out in all sorts of fantastic costumes—feathers, skins, &c., to make themselves look as fierce and savage as possible. According to the customs of the guild of porters, the Kiringosis have many peculiar prerogatives. They are looked up to by the men as their advisers on all questions affecting the caravan. They form a council empowered to consider all orders the leader may have to make. They take upon themselves the arguing of all grievances, e. g. whether the marches are too long, halts too seldom, or rations not sufficient in quantity or quality. Whatever be their decision, it is implicitly followed by the entire caravan. The great point therefore with all white travellers should be to get an influence over these men. It is only in this way they can hope to get along pleasantly.

Each of our Kiringosis was supplied with a flowing robe of crimson-coloured blanketing known as joho.

When all told we numbered 150 men, including a few engaged at Dar-es-Salaam, to proceed only a part of the way.

The presence of Dr. Kirk was of immense use in expediting our preparations. He being with us, no one dared to place obstacles in our way, and the people all knew very well that he had too extensive a knowledge of the country to be deluded by lies. Thus in three days our arrangements were completed, and we were ready to start. A Swahili guide had been engaged to take us to Ubena, and our five donkeys had arrived safely in a dhow from Zanzibar.

Every arrangement had, up to this time, gone so smoothly, and we had met so few of the irritating troubles and delays that so frequently, indeed, almost constantly, befall the European traveller about to start for the interior, that we could not but look forward, full of hope, to our ultimate success. When we looked at the fine selection of men, with Chuma at their head, full of tact and enthusiasm and experience, we felt ready to face any difficulty, and sure to master it.

Dr. Kirk has expressed the opinion that "no better organized caravan ever left the coast for the interior." For this no inconsiderable share of credit is due to himself—experienced African traveller as he is. So thoroughly indeed did Dr. Kirk identify himself with the Expedition, that he wished it had been in his power to have joined it. But undoubtedly it was the admirable care and forethought of Mr. Johnston which rendered the entire organization so complete. He made the subject a special study, and now, writing after the Expedition has completed its work, I can hardly recall to memory a single necessary article which we did not possess. Among the thousand and one minutiæ of a caravan's requirements such omissions are pardonable. Yet we were fully furnished in every respect when we started, and in so far as organization and wise equipment could secure the success of our undertaking, full credit is to be given to Mr. Johnston. Starting as we did with such manifest advantages, if our enterprise had not been a success, it could have only been through stupid blundering or the encountering of obstacles absolutely insuperable.

On the 19th of May, Dar-es-Salaam was a scene of the wildest excitement. With palpitating hearts we were up with dawn. The last article was packed, and everything got into marching trim. We were about to leave stone walls and comfortable rooms behind us, and make our tents our homes for an indefinite period. The men were mustered by beat of drum, and by the blowing of horns. Bales, bags, and boxes were turned out of the store. Some men with the reckless want of order so characteristic of the native porter, were found to be absent, and flying companies were forthwith hurried off to hunt up the absentees. Meanwhile the donkeys were brought out and loaded, and by 9.30 we were ready.

But at the last moment the guide was not to be found. After a good deal of hurrying to and fro, it was ascertained that he had gone off to a neighbouring village, and would not be back till after ten. We patiently waited, and after an hour's delay he honoured us with his presence. Order was then called, and each man fell into his allotted place.

First came the vanguard, lightly loaded with tents and cooking utensils, accompanied by a long mahogany collapsable boat, generously presented to the Expedition by its inventor, Admiral McDonald. The "Agnes," as I subsequently called it, after one worthy of all honour, was carried in front by two giants, and by its formidable appearance struck awe into the hearts of the natives, who imagined it to be a white man's cannon. Next came the main body, headed by the drummer, fantastically dressed, and proud of his soul-inspiring instrument, which, indeed, was a great and highly important adjunct of our caravan. Following him were the Kiringosis, clothed in all the savage grandeur of feather head-dresses and crimson robes. One of them carried a Union Jack, while the special loads of the others were the wire and the small boxes slung on the ends of poles which they balanced on their bare shoulders. The rank and file carried the bales of cloth and miscellaneous articles, some on their head, others on their shoulders, none on their back. The rear was brought up by five terrible nuisances—the donkeys with their drivers—accompanied by the headmen and Johnston, the former acting on the march as a sort of light infantry to look after stragglers, help the sick, and keep their eyes open for cases of desertion.

The signal was at last given to start. The drum beat its monotonous rum-tee-tum. The plaintive pleasant notes of the barghumi echoed and re-echoed from afar. Crack, crack, went gun after gun from porter or onlooker. The men with lusty shouts laid hold of their loads as if they were treasures, and then, with a sonorous recitative from the Kiringosis, and answering chorus from the men, they commenced their long march. I shook Dr. Kirk's hand and bade him farewell with a quivering lip, yet with heart full of great hopes and expectations. It was a moment when a little shrinking was pardonable as the dread uncertainty of the future rose before us. We were entering a valley of the shadow of death into which many have passed and from which few have returned.

In that great moment vivid pictures flitted

before my excited brain. I thought of Stanley beleaguered by thousands of bloodthirsty natives, and compelled to slay on every side to save himself. Would we have to do likewise? I pictured Livingstone dying in the swamps of Bangweolo. Might such not be our own fate? These and many similar speculations passed with lightning rapidity. But, in the proud consciousness of having started a great work, all doubts and fears were stifled in the bud, and waving answer to the hearty "Kwaheri, bwana mdogo!" of the Arabs, I turned my back on the Indian Ocean, and set my face resolutely for the interior.

Rounding the western side of the creek, our long winding line of men, stretching over nearly a quarter of a mile of ground, gradually emerged from the cocoanut plantations and cultivated fields of Dar-es-Salaam, and struck S.S.W. over a sandy, generally unbroken country. The dense impenetrable matted scrub and bush, which so often takes the place of forest in East Africa, here held possession of the land. In this tangled mass of vegetation individual plants could not be distinguished. The general separation of the whole into shrubs, creepers, and grass, was the most that could be done without a detailed examination, and it seemed as if one might with ease walk on the top of it without sinking down, so solid did it appear.

Along a narrow path we pushed our way, warding off the blows and scratches of the over-hanging branches as best we might. After crossing a small rivulet and ascending the bank on the opposite side we reached the little Swahili village of Pongwe. The moment we arrived the Kiringosis threw down their loads, and forming a ring round them they executed a break-down full of the most grotesque gestures, while the drums were thumped and the horns blown most lustily. This continued till the last man entered the camp.

In a few minutes our tents were erected under the spreading branches of a large tree. An ineffectual attempt was made to bring down a large green snake which had taken up its residence overhead, and we had to settle down with the pleasant consciousness of danger in our immediate neighbourhood. The bales, boxes, and other goods were built up in a heap, round which the headmen settled themselves as guardians. Pice were distributed to the men to buy food. For, be it understood, money is gradually being introduced along the coast, and there the natives are gradually beginning to understand its uses and prefer it to the old system of barter. Unfortunately this important innovation has not extended beyond the immediate vicinity of the coast.

These preliminaries over, we had the supreme pleasure of a good wash after our hard day's work. Then in the cool evening air, under the grateful shade of the tree, we had our dinner amid all the romantic conditions of African travelling, and over our cups of coffee we congratulated ourselves on the successful start we had made.

At sunset, according to custom, our caravan band, consisting of a native drum, a zomiri, and a barghumi or antelope's horn, struck up a native air. As the sun disappeared the flag was hauled down, and each one settled himself to the enjoyment of the evening. Our diaries were written up and our plans for the future discussed, and finally we turned in at an early hour. So ended our first day's march towards Lake Nyassa.

Pongwe was the last village on our route inhabited by Waswahili. By Waswahili I mean simply freed slaves or the descendants of such, gathered from every district of East Central Africa, including the countries lying around and between Nyassa, Tanganyika, and Victoria Nyanza. There is no doubt that in times previous to the Arab occupation of Zanzibar, a distinct tribe named Swahili existed. But, with the domination of the Arabs and the tremendous influx of slaves, they seem to have entirely disappeared as a tribe, leaving behind them only their dialect and certain customs.

It has hitherto been the usual fashion to speak of the Waswahili as the descendants of Arabs and natives—a hybrid race. But this I am convinced is not correct. However republican an Arab may be in associating with negroes, he has an intense pride of race. No matter how much negro blood there may be in his veins, yet if he has a trace of the Arab in him at all, he sticks most tenaciously to that side of his descent, and looks upon himself as an Arab, and widely separated from the negro. But as such half-castes are looked down upon by the pure Arabs, that is to say, the Arabs with unmixed blood, they are marked off as the Arabu wa mlima, or Coast Arabs. This, then, is the distinction between the hitherto rather conflicting names of Wa-mlima and Wa-swahili. The former are half-caste Arabs; the latter, descendants of freed slaves.

The Wa-swahili, having been formerly gathered from so many different districts, cannot be expected to show any marked tribal distinction, beyond the language and a few customs. They comprise every type of negroid physiognomy to be found in East Central Africa, from well-made faces, to others with prognathous jaws, thick everted lips, and bridgeless noses. In some instances hair on the face is found, though in the majority it is, as with most negroes, absent. From this heterogeneous tribe we drew the strange collection of men which made up our caravan—men who, while being called Wa-swahili, rejoiced also in the proud title of Wangwana, or freed men; rejecting with disdain the appellation of Pagazi, or porters. Such a servile name they leave to Wa-nyamwesi, from Unyanyembe.

Pongwe, the village at which we were now camped is an insignificant place, formed of a few quadrangular huts with overhanging eaves. It is clean and nicely situated, but presents no special peculiarities.

Next morning, as the grey dawn began to lighten up the landscape, the *réveillé* was beat on the drums, and every one was stirring in a moment. While the men selected their bales, and the donkeys got loaded, Johnston and I had a cup of coffee. Previous to starting the roll was again called, in case there might have been desertion during the night. Happily every one answered to his name, and forward we set once more.

After an hour's march across a slightly-undulating country, we reached the deep and unfordable stream of Mzinga, which flows into Dar-es-Salaam harbour. Here we made our first acquaintance with the customary African bridge. Nature in uncivilized lands is ever considerate, and generally makes up for the want of invention in the savage by providing something to hand ready for use, whether it be in food, in clothes, or necessary domestic utensils. So, also, in the case of unfordable streams, there generally stands some giant tree beside the bank, which requires but to be felled at its base, to bridge the otherwise impassable, and join the opposite banks. Such a bridge over the Mzinga we found ready for us, and though not the best footing for nervous individuals, yet, with a little assistance, it was quite passable.

The donkeys, however, not accustomed to such precarious pathways, could not be trusted upon it, and they required to be unloaded. The first one which happened to be standing on the bank was taken by surprise, and before it could take up an attitude of obstinate resistance, it was hurled headlong into the stream, and thus dragged across by an irresistible rush of porters. The others, however, seeing the ignominious fate of their comrade, indignantly prepared to make a stand. Each one laid its ears back, and put its tail between its legs, and defied the efforts of the porters to "run it in," kicking out right and left in the most alarming yet enjoyable manner. Finally ropes had to be resorted to, and all opposition was speedily overcome. After an hour's severe struggle, the donkeys were all landed on the opposite side, and reloaded amid the acclamations of the men.

Keeping our way S.S.W. along the valley of the Mzinga we passed a beautiful circular pool with water-lilies in the centre of a grove of trees. Such ponds, as I have already observed, occur frequently on the lower of the raised beaches which skirt the coast. They have evidently been hollows in the coral when under the sea, and they have remained as ponds on the elevation of the land.

Before camping we made a considerable ascent to a second and more ancient raised beach. In the upheaval of the coast there has been a marked pause; and thus two separate beaches have been formed; whereas, if the upheaval had been continuous and uniform in its rate, there would have been no such distinction.

As rain threatened us we were compelled to camp

early in the day at a small collection of huts, where the men could get no shelter. But this was a matter of little concern to them. In half an hour an entire village was run up with surprising dexterity. The men had already separated themselves into messes, for mutual convenience and comfort, and by the division and organization of their labours the most remarkable results were achieved. On arriving in camp each man in a particular mess or "khambi" knew exactly what to do. One as cook got out his pot, made a fire, and commenced boiling the water for the rice or other cereal; a second went foraging for the food; others, again, when there was no shelter in the villages, commenced building a hut or shed. Of these, one man would prepare the ground; a second bring poles; a third grass for thatching; while numbers four and five would do the building. The grass for beds would then be laid in. By the time this was finished the mess meal was ready, and they could sit down under cover and enjoy it in defiance of wind and rain. The rapidity with which a whole camp was run up was really marvellous.

The rain which had threatened did come on, but we were by that time prepared for it. Fortunately it did not last long, and the men were thus able to have a lively dance before night set in. The elasticity of spirits of Zanzibar porters is a marked feature in their character. At the end of a march they will come dragging along as if thoroughly exhausted, and ready to succumb to fatigue, and in an hour afterwards they are up and dancing like demons. There are no "Roger de Coverley" movements, or "Charing Cross crawls" there, but downright hard work, in which every muscle is brought into play. The feet stamp into the ground, like the hoofs of circus-horses, and arms and legs are thrown about in that alarming manner only to be seen in Parisian dancing-gardens. The ground becomes literally ploughed up, while shouts and recitatives are kept lustily going, and time is beat by clapping hands. The exertions of an ordinary working week seem to be comprised in three hours of such breakdowns.

Recommencing our march, we crossed a number of swampy hollows, through which we splashed in liquid mud. Gaiters I found to be of great use in such places for keeping the mud out of my boots. Owing to the previous night's rain, the grass was wet; so in pushing through it we enjoyed an involuntary shower bath, which became a vapour one as the sun rose and evaporation commenced.

We had capital fun in crossing the Mzinga stream a second time; in this instance without the aid of a fallen tree. I was in front, and wanting to show the men that I was no milksop, though I might be a greenhorn, I marched bravely into the stream. A few seconds after I was hauled out, spluttering and half drowned. It was too much for me. One of the men then tried another place. He got in a few steps all right, then suddenly went down overhead. With the most admirable presence of mind he kept his balance and hold of his load of cloth, which happened to be on his head. The men behind ran and got hold of it. He rose to the surface blowing like a whale, but he had saved the cloth. As the stream proved to be fordable, all except the middle, the tallest men were selected, and stationed themselves up to their necks in water, and then with a sonorous refrain the bales were passed safely overhead in the most animated manner. The men enjoyed these occurrences with heart and soul.

Everything having been safely taken across, and the donkeys treated in the ignominious manner already described, the episode was consummated by the men shouldering Johnston and passing him like a bale to the opposite side, when all the rest of the men gathered round, and, forming a jolly procession, carried him several hundred yards. There was a good deal of fun in observing the struggles of those who could not swim to get over. A derisive crowd watched each attempt, greeting with cheers the successful, but covering with ridicule those who floundered, and only giving a helping hand when their situation looked serious.

Before camping heavy rain came on again, which by no means added to our comfort. The occurrence of rain at that time rather took us by surprise. We had calculated when we had started that it was all over, but now it seemed to have set in anew in a most unusual manner, filling the streams to overflowing, forming miles of swamps, and by alternate showers and warm sunshine making the atmosphere exceptionally malarious. The most dangerous part of the whole year for fevers is that immediately following the rains, when the submerging waters begin to subside, and the sun to assert itself through the overhanging clouds, raising the most fetid odours from the putrescent vegetation, which poisons the water with its gases. And yet, owing to the alternations of the climate during the year, this is the time when all caravans must start for the interior if they want to get any distance inland before the next rainy season comes on—more especially as that season commences much earlier in the interior than it does on the coast. This period, notwithstanding its dangerous drawbacks, has also its advantages. The days are generally cloudy, and the temperature is at its lowest, so that the hardship of travelling under a blazing hot sun and oven-like atmosphere are not experienced till the traveller has become somewhat inured to the work.

On the fifth day of our march from Dar-es-Salaam we once more descended from the upper to the lower raised beach, camping after an exceptionally long march at the large village of Liwali. This day's tramping rather tested the powers of the men, and showed who were the best. Many of them had to be assisted with their loads into camp as much as two hours after the foremost.

The following day being Sunday we naturally formed the resolve to keep it as a day of rest. The men, dressed in their best, were mustered, and the roll called, with no one amissing. At Liwali we found three villanous-looking Arabs staying, who had evidently fled from the clutches of their creditors on the coast. The village or villages form a collection of huts dotted irregularly about a magnificent plain, with large rice-fields growing in rich soil deposited from the Mzinga stream, which here occurs still on our line of route. If I am right in considering this plain to be the lower of the raised beaches, it would here appear to be nearly thirty miles broad. In the afternoon we wrote letters home, which were to be despatched by the brother of our guide, who had accompanied us so far on our way.

Recruited by our halt we recommenced our journey, though the rain was falling in torrents. Like somewhat "green" travellers we tried the umbrellas we had brought, but these soon were in shreds and tatters as we ran tilt against thorny branches in the narrow pathways. The country was almost one entire swamp, and where not so the footpaths were cut into deep ruts by the heavy rains, which made it a weariness to the flesh to flounder along.

The country after leaving Liwali became much more picturesque and varied in its character. Here and there it was cut into small hills or ridges by the numerous rivulets which form the upper waters of the Mzinga. At the village of Kikonga, picturesquely situated on the top of one of these small hills, we camped on the 26th. Here we had a fine view of the Liwali plain, extending away to the sea.

Kikonga was the best of the villages we had yet seen. Embosomed in a grove of cocoanuts, it formed a circle which occupied the entire area of the hill top. The houses were unusually large. They were quadrangular in shape, with the ordinary thatched roof. Occupying one half of the front of the house was a slightly raised portion, called the baraza, open on two sides, and roofed over for shade. Here the village gossips collect and retail their latest news. Here all business is transacted and friends meet friends.

The baraza is open to all who may want to rest, and the porters of a caravan take up their quarters there without asking permission. The long, projecting eaves of the roof hang out so far as almost to hide the walls from sight, warding off the scorching blaze of the sun, and thus ever keeping the interior cool even in the height of summer. The only light admitted is by the doorway, which being small does not much assist in illuminating the interior. But the deep shade which obtains there is extremely grateful to the traveller after tramping for hours in the fierce glare of the sun. It is soothing and refreshing in the extreme to the passing wayfarer, however unsuited it may be for a prolonged residence.

The people themselves, of whom I have not yet

said anything, have nothing special in their appearance to mark them off from the porters of the caravan, except in the matter of clothing, which we notice is diminishing in a rather alarming manner. The men and women, however, have a loin cloth each, and the latter even have a small detached square of cotton across their breasts. They all have small features, and the long narrow cranium so characteristic of the negro.

There is a variety of fashions in hair-dressing. The style of the Zanzibar women has gained a footing, and will no doubt soon supersede the more "vulgar" styles of the ordinary Washenzi. The head when dressed in this mode exhibits rows of short platted frizzly hair, extending from the nape of the neck to the brow, and thus has a furrowed appearance. The less ambitious *belles* part their hair in the middle, and work it into long strings with grease, which are frequently terminated by balls of mud. Others, again, work the hair into horn-like projections. Ornaments do not appear to be much in vogue.

One most noteworthy fact is the apparent prolific qualities of the Wazaramo women. It is quite a rare occurrence to see one without a child in her arms. In this respect they present a great contrast to the Waswahili women, who are rarely seen with children. This latter fact is due mainly to two causes; first, the dreadful immorality prevalent among them; and secondly, the use of drugs to produce abortion, as children are generally looked upon as a nuisance.

A very noticeable circumstance was the general absence of weapons among the men. Neither arrows nor spears were to be seen, and there was an aspect of security and peace reigning, which, now that I review the tribes I have passed through, appears to be quite unique. There seemed to be an entire absence of wars or tribal feuds. That may be accounted for by the fact that on the one side the only people were the Wakhutu, a weak, timid race, incapable of doing harm; while on the other side there were the coast people, who, though infinitely superior to them in every way, were kept in check by the authority of the Sultan. To this subject, however, I shall have to recur further on in the narrative.

Leaving Kikonga, we had a short march to Kidokwe, where we were compelled to camp early, as the next village, Mkamba, was too far to be reached in one day. At this latter place we arrived on the 28th of May.

In a village some distance from Kidokwe, we were the spectators of a curious scene. In the square of the village, propped against a tree, sat a poor woman, apparently half dead with illness of some sort, and looking very much as if she was in the stocks. Round about were some huge pots of native beer, presided over by a witch-like old beldam, who ever and anon ladled forth the sour muddy mixture to whatever thirsty mortal might ask for it. On one side was a great array of drums of all sizes, notes, and tones, from a brass cymbal to the bole of a huge tree, hollowed out, and covered over at the ends with skin. These were thumped with all the vigour that brawny arms were capable of, reminding one of the typical village blacksmith. The noise produced could not have been equalled by any iron foundry. Round these instruments was a half-drunken crowd, whose wearied looks spoke of a want of the previous night's sleep. In front of the woman the men danced in succession, with movements and gestures the most extraordinary, though evidently not in any way obscene.

On inquiring the meaning of all this, I learnt that they were employed casting out devils from the woman in front of them, and to do this they required to use the most powerful charms they could think of, namely beer, dancing, and music. They had now kept this process going on for about twenty-four hours, and had already beguiled a few of the devils out of her, although, to my thinking, "the last state of that woman was worse than the first." The struggle must have been a severe one; but who could withstand such jolly physicians, who plied the patients with merry-makings, beer, dancing, and music?

The devil, according to native belief, is peculiarly amenable to the charms of the drum. Whenever any one is ill, and consequently supposed to be possessed of the Evil One, he is treated to an Ngomma, or drumming. As soon as the devil, captivated by the so-called music, has his usual 'cuteness dulled, he is enticed by a medicine-man into a stool, or some other article ready for his reception. He is then carried about from place to place to the sound of his favourite instrument, till he is so completely bewildered that he is unable to find out his former victim, and consequently decamps in disgust. If such is the music which soothes the disturbed spirit of the Evil One, it must surely be one of the peculiar tortures of wicked musicians to be invited to his majesty's concerts!

CHAPTER IV.

MKAMBA TO BEHOBEHO.

THE important village, or perhaps more appropriately the district, of Mkamba, lies S.S.W. of Dar-es-Salaam, at a distance of thirty-five miles as the crow flies.

The soil is extremely fertile, and well cultivated; it yields all the varied products of the east coast, and supports a large population of well-to-do natives of mixed Wazaramo and Waswahili. The country, compared with the monotonous tracts we had just passed, is charming in its broken outlines, its wooded ridges, and well cultivated hollows. Here and there are clumps of palms, mixed with mango, jack-fruit, orange, guava, and papaw; while on all hands plots of cassava, millet, Indian corn, rice, sweet potatoes, ground-nuts, and beans abound. Numerous villages shadowed in deep groves are revealed by the curling smoke.

Everywhere the people are seen moving about, for the time of our visit is the busy season, when the cereals are ripening, and require extra vigilance to protect them from the spoliation of birds and monkeys. Hence cries, which might have been startling to the ignorant stranger, resound on all sides.

No one who observes the negro at this time of the year would for a moment dub him an idler. His days are given to hard work in the fields, and the night air rings with his joyous light-hearted glee. In anticipation of a good harvest he can afford a "spree" with his hoarded grain from last year's produce; so the beer is brewed, and the energetic dance and joyous song tell of an exuberant delight devoid of care.

At this place, with these surroundings, we had now arrived. Our goods were stored from the rain under the verandah of the chief's huge quadrangular house, and our tents pitched in the square. The chief, an extremely pleasant and hospitable little man, dressed in the shirt-like kanzu of the coast, received us with much genuine good-will, which we could not but appreciate. To do full honour to such distinguished guests he celebrated the event in the evening with unlimited pombe (native beer) and a dance. Acting under the idea "the more music the more honour," the dancers tripped the green to the inspiring clamour of twelve drums of different sizes and tones, from one a foot in length and half-a-foot in breadth, to others three feet in length, which were simultaneously thumped *con amore*. The dancers were a mixed company of the sexes, dressed in their best.

They commenced in a very decorous and respectable fashion, with slow movements, and much bowing and scraping, while a tune was chanted of a very solemn character, which made us hope that after all it would be a performance we might be able to sit through. We were, however, mistaken. A change came over the spirit of the scene. The beating of the numerous drums became faster, and the performers indulged in curious freaks and startling effects. The zomiri player screeched louder, and blew till his cheeks seemed as if they would burst, while he swayed himself to and fro as if to assist in producing the necessary amount of wind. The singing grew more and more energetic. Clapping of hands commenced, and occasionally a shrill scream would ring through the air. Arms and legs were thrown about with unpardonable levity, and a terrible whirlpool of human beings surged about in the maddest excitement. With tenfold energy the drums were belaboured. The eyes of the zomiri players seemed starting from their sockets. The ear-piercing screams rang more and more shrill, and wilder and fiercer grew the dance, till individual figures could hardly be traced. The sun set, twilight passed, and darkness reigned, but the dance proceeded with unabated energy. The ruddy glare of a large fire gave renewed effect to the scene, until we could almost imagine ourselves in the infernal regions and witnessing a witch's carnival.

This devil's dance continued all night, producing

nightmares and awful dreams as we now and then dozed off amidst the infernal uproar. Though such orgies are awful to the civilized ear when in the immediate vicinity, yet I must own that during the night, when about a mile off, I felt a certain piquant enjoyment in lying listening to the wild tumult of sound. The weird thrilling scream of the women, and the wild clangour of the drums could not but affect a lively imagination.

As we had now arrived at the confines of the well cultivated and populous districts, it was necessary to be cautious in our movements, not pushing too hastily forward, but taking as our motto "be sure of every step before making it." As our guide was not certain about the best route for such a large caravan, so as to get food, we decided to send him forward a few marches to report upon the country. In the interval we enjoyed ourselves after various fashions; Johnston kept up incessant inquiries regarding the countries we would pass through, and spent the time in other profitable ways, while I wandered about hunting for beetles and butterflies, beasts, and crawling things of different kinds. But though I searched most assiduously the result was invariably disappointing. Everything animate seemed determined to keep out of my sight. Rarely did any living creature fall into my snares. Still, a lover of nature always finds something in these rambles interesting enough to draw him out. The great stillness of the forest and the entire novelty of his surroundings are of themselves sufficient to keep him from wearying.

But many pleasing episodes frequently occurred, which gave food for reflection and glimpses of the rudimental mind of the savage. Hidden in the forest, we stand, perhaps, and watch the ways and customs of the villagers. We note the domestic scenes and their ways of dealing with each other. Arriving suddenly among them, we smile to see the startling effect upon their superstitious minds. We meet a girl in the forest walking in a narrow footpath, and though we are known to be harmless, yet fear and awe are depicted in her face. She steps back into the bush as far as possible, and hardly dares lift her eyes from the ground, except with the frightened side glance so characteristic of an ill-used dog, which expects another lash and yet knows that it dare not resist. We speak to her, and she seems to shrink within herself. To a lesser extent it is the same with the men also. Caught alone, they cannot resist an inward tremor as they recall the strange and unearthly stories told of the mysterious white man. Only when together in numbers do they appear at all easy in our presence.

One day, having nothing to do, we were suddenly taken with a fever for hunting. Wonderful stories had been told us of the abundance of hippos in a small lake a few miles distant, which did great damage to the crops. We resolved to become the benefactors of the natives and rid them of their ravagers.

Fired with the idea we at once seized our heavy rifles, and taking one or two men and a guide set off for the field of action.

The men wanted to carry our rifles for us, but this we loftily refused as derogatory to the dignity of sportsmen. For the same reason we refused to be borne across a stream, and I dashing in got a fine ducking, by tripping over a root in my haste to show that I could rough it with impunity. I, however, only laughed, and declared it to be the very best of fun. Pushing along for an hour, we were overtaken with rain. The path disappeared, and the grass became drenched with wet. Such little troubles were of course beneath the notice of sportsmen! Then we entered a horrid miry swamp, full of pitfalls and holes, which squirted mud over us till we looked the most blotched of individuals. We slipped and floundered about in the most wearisome manner, till at last, finding my rifle becoming rather heavy, I argued to myself that it would look more dignified if an attendant carried it. I should then be able to speak of "my gun-bearer!" This "happy thought" I at once acted upon, and Johnston, glad of such an example, also relieved himself of his gun. Another hour through swamp, jungle, and forest, and we emerged finally beside the so-called lake, followed by a crowd who were eager to witness our exploits, and whom we of course determined to astonish.

But where was the lake? A stretch of marshy-looking country spread out before us; and it was only after some investigation we made out that a body of water did lie at our feet, but hidden by a thick covering of floating vegetation, except at the centre, where we were told the hippos were to be found. This took us rather aback, as we would require to wade an indefinite distance by a hippos' track to reach the clear water. Our fervour for sport rather cooled at this prospect. Might there not be crocodiles or water-snakes hidden beneath the vegetation, ready to take a mean advantage and snap at us unseen?

We resorted to various plans in order to get a sight of the monsters. Trees were climbed and the treacherous edge of the mere was patiently investigated, but all to no purpose. The natives began to look disappointed, and evidently we were fast falling in their estimation. This rather nettled us. The idea of an ignominious retreat was anything but acceptable. Still the attempt to reach the open water seemed to mean an amount of trouble and discomfort out of all proportion to the value of the possible sport.

Chopfallen, we were therefore about to retire to our camp, when from the haunt of the hippos came three distinct grunts. We were fairly electrified. Were we to brook these notes of satiric triumph, and meekly accept our defeat? It was impossible! At once boots and coats were off, and in a minute Johnston dashed into the open track through the floating vegetation.

Struck with admiration, I stood and watched him. His footing seemed to be rather shaky from the way in which he moved forward, with "light springing footsteps." Further and further he boldly ventured. Deeper and deeper he got, while ever and anon the hippos grunted out defiance, and the mob shouted encouragement on the banks. The liquid mud reached his hip, then his waist, and gradually crept up till his armpits were reached, and still he had not attained the open water. It was a critical moment. He paused and looked back; then sternly making another step forward, he suddenly dropped out of sight, with only his gun above water. A few minutes later, he was hauled to the bank, covered with a thick integument of odoriferous mud, baffled and defeated; and loud rose the victorious grunt of the hippos.

While the men scraped off the obnoxious cuticle, I made an attempt in another direction, and returned in like manner to take my turn under the scraper. Johnston the while stood shivering and chilly. Feeling as if it would relieve us to indulge in a little strong language we commenced our return, crest-fallen, and in a most pitiable plight, though before we once more re-entered Mkamba the pouring rain had pretty well washed the mud out of our clothes, and saved our washerman that trouble. We concluded that night that hippopotamus hunting was not exhilarating.

On the following morning, Johnston felt a pain

in his back which he supposed to be rheumatic, and under that belief, took what he understood to be appropriate medicines. All too late, however, he found out that it was the commencement of a more deadly disease, as we shall see, in the course of our narrative.

Previous to this excursion, some capital observations were made, which fixed the latitude of Mkamba. From the very first day's march, Johnston had assiduously attempted to take meridian altitudes of various stars, but owing to the cloudy condition of the skies, had failed on every occasion.

On the third day after his departure, our guide returned from his reconnaissance of the country ahead, and brought a very unfavourable report. There would be eight days' march, in which no food could be got. Supplies would therefore require to be provided at Mkamba and a village two days further on. To this task, then, we immediately set ourselves. Men were despatched to every village to buy rice. In an hour, the camp was deserted, and remained so till the evening, when they returned, laden with bags of food.

On the following day, all the wooden mortars of the place were secured for husking the rice, and round these, animated groups pounded vigorously, keeping time to most peculiar chants. These mortars or "kinus" are wineglass-shaped blocks of wood, hollowed out to receive the rice, which is belaboured with a pole four feet long, struck vertically downwards. The process is very slow and laborious. The husks are removed by the horizontal motion of an almost flat basket, which separates the heavy grain from its light covering.

Having occasion here to flog one of the men for flagrant disobedience, I was greatly disgusted to see how much the others enjoyed the sight of the man's punishment. Yelling with laughter, they seized him with savage glee, had him down in a minute, turned him on his face, and held him as with a vice, while the punishment proceeded under the cane of Chuma.

There is, however, one good trait in the men's character. They never harbour any grudge or revengeful feeling if their punishment has been just. Half an hour after the ordeal they may be seen laughing and joking with the very men who held them. This of course may arise from the fact that no sense of degradation accompanies punishment, not even that of flogging; and being naturally light-hearted, and of a devil-may-care disposition, they soon forget.

While preparing for renewing our march, one of our porters disappeared, and suspecting that he had made direct for the coast, we despatched one of the headmen and a porter to hunt him up.

After this unpleasant detention, we with pleasure got once more *en route*, though Mr. Johnston was much pained by his illness; and the rain, which had never ceased during our stay, still continued with unabated violence, submerging all the low-lying tracts, and turning the footpaths in the higher ground into raging torrents, along which it was a weariness to struggle. Still these were pleasant days to us, when with robust frames and eager enthusiasm we only rejoiced in the troubles and hardships of our march. We had come prepared for all this; indeed, we should almost have felt disappointed if our route had proved easy and pleasant. Ridiculous as it may seem, we thought ourselves entirely unworthy of the honourable title of African travellers until we should have undergone such an apprenticeship of endurance and physical discipline. Disease and bad food had not then broken our spirits and undermined our constitutions. But that time came only too soon, and then one of us was added to a roll which in the sanguine hopefulness of good health we did not anticipate.

At Madodo, where we camped the first night after leaving Mkamba, we were astonished at the reappearance of our lost porter. He gave a most creditable account of himself. Hearing at Mkamba through some unknown source that his wife was very ill on the coast, he clandestinely went off to see her, fearing that, if we knew, he would not be allowed to go. In three days he traversed more than 120 miles, saw his wife, and returned like a faithful fellow to his work. Such a deed is worthy of honour in any land, but is especially noteworthy among such a race as the reviled Waswahili. During the night the men sent after the runaway returned, having followed him to the coast, marching night and day. There they heard the true version of the affair, and came back to find him safe in our camp.

Two more days' toil under the afore-mentioned wretched conditions brought us to the village of Msangapwani, where, as we had now reached the borders of the inhabited and populated district, we were compelled to halt to collect more food in anticipation of the desert marches.

When we arrived at Msangapwani neither of us had a single dry article of clothing, so little had the sun shone to give us an opportunity of freeing them from moisture. As one suit got wet another was put on, till at last we had to be content with a wet one. Even our blankets were damp and clammy, while the close steaming air was impregnated with malaria. The natural consequence of this was that Johnston became much worse. Here also he discovered, what he had not hitherto suspected, that it was dysentery, and not rheumatism with which he was troubled. Meanwhile I was likewise laid low with an attack of fever.

In these circumstances matters looked rather lugubrious and melancholy. We tried to joke feebly with each other on our ailments; but I could hardly hold my head erect, and Johnston was looking the agonies he would not express in words.

While the men were employed collecting and preparing food, we resolved, after a council of war, to dismiss the guide we engaged at Zanzibar, as he clearly knew nothing about the road.

He, however, saved us that trouble. Having got several rupees from Johnston, to go and buy rice from a man who was wanting, as he said, to dispose of it for money, he disappeared, and was never more heard of. We, however, were quits. The money we had advanced at Dar-es-Salaam just paid him, according to the "rate," for the time he had been with us; so neither lost, and there was little difficulty in securing another more competent to conduct us.

We had here a visit from an Mganga, or medicine man, fantastically dressed. He had with him an obscene image of a woman, clothed with beads, and looking like an absurd toy. Though he did not appear to have very much respect for his goddess, if such we might call it, yet he would not sell it under an exorbitant price. When using it, he irreverently shakes it in a bag, and, thus awakened, the oracle speaks, unheard, of course, by the materialistic ear of the mob, and declares its mind on whatever subject it may be consulted.

After a three days' enforced stay at Msanga-pwani, we once more got under weigh. We both presented a very pitiable and woe-begone aspect when we stepped out. We were by no means promising-looking leaders. As usual, I was in front, trying, like a drunken man, to assume some dignity of appearance, though I have no doubt the attempt was rather comical. Johnston was in the rear, even in a worse condition. Our march was, as before, made miserable by drenching rains, and we had to struggle successively through long grass, swamps, and deep swollen streams. At midday our efforts ended in a complete collapse. Johnston arrived an hour after me. As he appeared, I tried to look jolly, and to hail him with a consolatory remark, but his only response to my weak attempt was a groan, as he sank exhausted on the ground.

Two hours' rest and a cup of hot tea somewhat reinvigorated us, and as we could not camp where we were, we staggered on a little further to a more suitable locality in the forest. The tents were pitched, and a boma, or thorn fence formed, inside which the men made their huts. For three days we were confined in our tents *hors de combat,* and unable to do anything. Chuma, however, was equal to the occasion, and kept everything in order. It is under such circumstances that the value of a man like Chuma is understood. One with less influence and tact would be unable to keep down riot and disorder. One with less honesty would certainly take the opportunity to help himself in various ways. The worst of my attack was soon over, and I recovered with remarkable rapidity. Johnston also decidedly improved.

Our position in the forest was a somewhat awkward one, distant as we were from all food supplies, and it became necessary to move forward. Owing to the rains, the incompetency of our former guide, and the scarcity of food in the country, our progress had been exceedingly slow, and we got quite irritable in our anxiety to push ahead. As soon therefore as Johnston found himself able to rise, though still unable to walk, he determined, in spite of his illness, to set out once more. This was an unfortunate decision. He was improving rapidly, and a few days' more rest would have given him a fair chance of throwing off the dysentery. But in his eagerness to proceed there could be no rest for him, at least until we should have reached an important village, called Behobeho, which we had heard much of. As we unhappily were not supplied with a hammock or other convenience for carrying an invalid, we set to work, and with the aid of some of the men, we contrived to fix up a rude concern, which was certainly not of the most comfortable nature, but was the best we could produce with the materials at our command.

Resuming our journey, we kept in a south-westerly direction, entering the drainage basin of the Rufiji. The streams we had crossed as far as Mkamba form the head waters and tributaries of the Mzinga River, which, as we have already noticed, flows in a northerly direction to the Dar-es-Salaam creek. Between Mkamba and Msangapwani four considerable streams, with a number of minor ones, find their way directly east to the coast. At the point we had now reached the versant and drainage is towards the river Rufiji.

The country maintained much of the general character which we have already described, only it was much more flat, spreading out like an immense plain. We missed the coast fruit-trees—the moisture-loving cocoa-nut, the luscious mango, and the stinking jack-fruit. No cultivated fields or inhabited villages met our eye. To add to the desolation of our surroundings, a great stillness pervaded the solitude, and nothing animate seemed to exist. There were no pretty chattering weaver-birds; no golden-vented thrush sent forth its joyous music; and the tooting of the tepe-tepe was hushed. Sunbirds, orioles, all were alike absent. Only occasionally from the surrounding forest was the cry of the hornbill, or of the omnipresent wood-pigeon heard, or it might be the caw of a parson crow.

In trudging along through some of the forest tracts, we were frequently in danger of broken legs, or sprained ankles, owing to the hundreds of deep grass-hidden pits from which the natives extract the gum-copal, of which all our best varnishes are made. As it was we several times got severe falls in very sudden and unexpected ways.

Having to occupy Mr. Johnston's place in the caravan during his illness, I began to learn what a dreadful nuisance the ordinary Unyamwesi donkey is. We had five of them bringing up the rear, each with a load equal to what would be carried by two men. Each one, however, required the services of a man to tug, swear at, and thrash him. So that a donkey was only equal after all to one man. But their value for purposes of transport sank into insignificance when one thought of the immense amount of trouble and delay they caused to the entire caravan. In a week the donkey-boys had lost all moral control of themselves, and indulged in nothing but profane language, till it was feared they would become insane from the amount of irritation they were subjected to. While loading in the morning it required about ten men to each beast, and success was only attained after a severe struggle. Once loaded they usually took the first opportunity to walk into the nearest thorny mass of scrub, where, madly and frequently successfully, they would strive to leave their loads behind. A quarter of an hour would then be spent in releasing them from their entanglements. Finally, all ready, off we would go, though with our moral balance very much upset. The donkey for a time would look like a lamb in its meekness, till, finding our vigilance relaxed, smash it would go against a tree-trunk, breaking girths and saddle-bags, and scattering the load on the pathway. Then, with unmitigated thumps and screams of rage, the donkey-boy, at his wit's end, would dance about shouting for help. Thus another half-hour would be wasted in mending the saddle-bags and reloading.

When a stream had to be crossed more trouble and delay ensued. The loads required to be taken off and carried across by the men. Then a grand fight would commence to get the donkeys over to be reloaded again. In this manner a day's march in charge of the perverse brutes, became a weariness to the flesh, and left us little time to attend to anything else.

It was with a genuine feeling of relief we saw these weak creatures die off by some mysterious malady. What exactly was the occasion of their death one after the other it would be difficult to say. I do not think it was the tsetse; neither was it bad treatment, nor the want of food. The climate and the nature of the food seemed to be the chief causes. In the low swampy regions, covered with dense jungle-grass, neither bullock, horse, nor ass seems to thrive, except where the greatest care is taken of them. They require to be either fed on specially collected food, or allowed to stray only on parts which have been under cultivation, where the grass appears to become more wholesome.

After my experience with donkeys, I would certainly advise no one to use them in an African caravan, either for transport purposes or for riding. In a country like East Africa, the use of donkeys for the latter purpose is to be specially avoided. The riding is destructive of all energy where energy is most needed. Indeed, I am convinced that the man who rides 300 miles on a donkey in these regions may be set aside as physically a bankrupt man, and incapable of doing any good work as an explorer. This may seem strange, and may require explanation.

It is a well-known fact that the only way to resist successfully the enervating effects of a humid tropical climate is by constant exertion, and by manfully fighting the baleful influence. The man who has nothing to do, or won't do what he has to do, is sure to succumb in a few months, and degenerate into an idiot or a baby. He becomes the helpless victim of manifold bilious troubles, and is continually open to attacks of fever, diarrhœa, or dysentery. His mental energy flies with his physical, till any sustained thought is impossible, and to pass the time he must dose night and day, except when he is grumbling and defaming the climate. Hard constant work is the great preserver. Sweat out the malaria and the germs of disease, and less will be heard of the energy-destroying climate of the tropics.

These facts have a very obvious significance in reference to the use of donkeys. The traveller who uses donkeys has seldom occasion to exert himself. He sits hour after hour under a pitiless sun, with a dreadful lethargy creeping over him as he moves forward at a snail's pace. There being no physical exertion, the necessary perspiring conditions are not obtained, and he feels shrivelled up into parchment. This might be avoided, of course, if he had the terribly hard work of keeping his donkey on the pathway, and making it get along. But that would entail more labour than if he walked on his own feet. So he provides himself with a donkey-boy, who does all the swearing and the thrashing.

Three or four hours under such conditions and the traveller reaches camp half dead. Concluding that his dreadful weariness is the inevitable result of the climate he walks feebly into his tent to have a nap to recover himself, whereas the only possible cure is to be up and doing, interesting himself in everything pertaining to the caravan, looking into this thing and the other, shaking up the idlers, and producing universal activity, till he is bathed in perspiration. To the indiscriminate and lavish use of donkeys may be ascribed a very large proportion of the troubles which have afflicted many caravans. Donkeys are doubtless a good thing when used judiciously, but then it requires a very strong-minded man to keep within due bounds.

Two marches from our camp in the forest brought us to the banks of the Rufiji River, at a village called Kimkumbi. These two long marches naturally told very severely upon Johnston, owing to the rude mode of conveyance at our command. The pain of his disease was much aggravated by the exceedingly unpleasant jolting trot of the carriers along slippery pathways, where a firm, steady footing was quite out of the question. On several occasions they nearly fell with their precious load. Moreover, to be carried under such a sun was of itself sufficient to knock up any one. His tortures under such circumstances were simply dreadful, and when we arrived at Kimkumbi he felt half dead. His body was stiff and swollen, and he was utterly unable to taste a particle of food. The only means by which he could sustain his waning energies was an occasional sip of brandy and water.

In order that he might recruit slightly, a day's halt was determined on. This I took advantage of to examine the river.

The Rufiji, notwithstanding its large size and apparent importance as a water-way to the interior, has as yet been little explored, owing to the difficulties attaching to its navigation, and the malarious nature of the bordering country. The sight which met my view was exceedingly disappointing. Instead of a noble river, winding along between well-defined banks, there seemed to be only a great swamp broken here and there by sandy islands, and huge sedgy tracts, the haunt of innumerable herons, storks, ducks, geese, king-fishers, ibises, and every variety of waterfowl to be found in the tropics, along with the clumsy hippopotamus and the dangerous crocodile.

On inquiry I found that the villagers had only two or three canoes, each able to carry not more than four persons with safety, and used solely for fishing purposes. Selecting one of the best I crossed the river to form some notion regarding its nature.

To judge by the eye, the Rufiji at Kimkumbi is half a mile in breadth. A considerable proportion of this is taken up by a large sand-bank in the centre. Along the northern side the current is swift and the water deep—though far from uniformly so—and snags and mud-banks occur in the most unexpected places, even where the current is greatest, and the water deepest. Along the opposite side also the river is both swift and deep.

The work of navigating the river even by small boats must be one of difficulty, as, owing to the muddy nature of the water, the snags and banks cannot be seen. Indeed our canoe ran aground several times from this very cause, though guided by a native who might have been expected to know where such existed. Doubtless it was their shifting nature which frequently baffled even his local knowledge.

We were told, however, that canoes capable of carrying twenty men frequently arrived from the coast with salt to barter for gum-copal and rubber, which are here collected in small quantities. I think that this is a rather doubtful story. But whatever may be said regarding the navigation of the river by canoes, I am convinced that no boat of dimensions large enough to do much trade can ever come so far as this village.

From Kimkumbi the river trends away E.S.E.,

while it comes from the S.W., and thus forms a considerable curve in its course. The natives on the banks do a good deal of fishing, capturing considerable quantities of fish with fleshy tentacles, which have been described by Burton as "tasting like animated mire." The larger fish are speared from canoes. These they smoke dry and trade with all over the country. Cotton is grown to some extent at this place.

Somewhat improved by the day's rest, Johnston, despite my anxious protestations, determined to start again. There should be no more stoppage till Behobeho was reached!

The conditions of travelling had now, however, very much changed. The rainy season was over, and from a clear cloudless sky the sun beat down with withering effect. The change in the appearance of the country was no less marked; swamps and marshes were replaced by dry, burnt-up deserts, which were extremely painful to traverse, as the mud, during the rains, had been cut up and wrinkled by the feet of wild game into a surface of the greatest irregularity, which had then got baked and hardened by the sun to the consistence of stone. Over this the men painfully limped with their bare feet. Not a drop of running water was to be got, and we had to be content with the slimy water of pits or small ponds, befouled by rotting vegetation. The dense matted bush and tall jungle-grass with which we have become acquainted in

Uzaramo, gave place to open ground covered with scattered thorny acacias. These proved to be a terrible nuisance to the bare feet and legs of the porters. The fallen thorns on the pathways were continually getting into their feet, and laming them in the most painful manner.

We were now compelled to take a route which led us away about W.N.W., in order to make a détour round a great uninhabited stretch, which at that period was almost completely submerged.

The crossing of the Rufiji with the frail canoes of the villagers would have been both a long and a dangerous process. So it was deemed advisable to take the longer but safer route by Behobeho.

The first day we marched for two hours along the banks of the Rufiji; then striking away from them we traversed a shrivelled up plain with small stunted trees, camping early at a village called Mtemere. We found the country bordering the river here was covered with deep lagoons and backwaters, where myriads of wading and other aquatic birds found a congenial residence. As the river could not be seen from the village I made an attempt to reach it in a most extraordinary log of hollowed wood, which rejoiced in the name of a canoe. It did not say much for the arboreal growths of the neighbourhood that no better tree could be got than one shaped like the letter S. Yet such was the elegant outline of the craft in which I took passage. However, in the quiet lagoons, I thought we could not be in much danger, and I squeezed myself in accordingly. We failed to reach the Rufiji, but none the less enjoyed the pleasant sail among the huge sedges, with their screaming feathered inhabitants flying about in immense and varied flocks.

The next three marches led us still in the same direction (W.N.W.), through the weary desert, so utterly devoid of all interest as to make the sight of a herd of antelopes, or other large game, a matter of the greatest excitement. At some times of the year this district must be overrun with game, to judge from the way in which the ground is ploughed up with their feet. When we passed, however, they seemed to have shifted to more congenial parts. Occasionally a herd of antelope or quagga appeared in the distance, but carefully keeping out of gunshot; and as the open character of the ground and the abundance of dangerous thorns made stalking impossible, we had to be content with watching them a good way off.

By this time I had forgotten the humiliating result of our Mkamba hunting, and having camped one day at what seemed a promising place, I went out, taking with me the giant Beduè, to knock over, as I expected, a few nice antelopes. With the sporting fever still as strong within me as ever, off I started, full of sanguine hopes. The day was fast declining. We moved stealthily about for some time, like villains intent on mischief, peering eagerly here and there, and straining eyes and ears. As the shadows deepened, our imagination conjured up abundance of game. Like wary sportsmen, down we would drop on our knees, and suppressing heroically any interjections which might be suggested by the probing of the numerous thorns, we would carefully crawl up behind a bush, only to find that after all there was nothing to be seen. Darkness come on, and our toil was still unrewarded. We began to think of returning home, when suddenly, on emerging from a dense bush, we came upon a fine group of large antelopes. We saw each other simultaneously, and we exhibited mutual surprise. I was so struck with the fine pose of the figures and their look of alarm and astonishment, that I utterly forgot to put my gun to my shoulder; while they, paralyzed with fear, stood for a moment, uncertain what to do. Bedué, more practically minded, finally called out, "Piga, piga bunduki, Bwana!" (Shoot, shoot, master.) The words instantly broke the spell. With one grand bound they were into the forest, and lost in the darkness before I could raise my gun. We did not get another chance; and fearful of losing our way in the darkness, we returned, not without difficulty, to camp, heated and tired by our exertion, and withal disappointed with African sport, the only result being a bad cold, torn clothes, scratched skin, and general depression of spirits.

Small snakes are very common in this dry region,

and we came frequently into unpleasant proximity. More than once did they glide over my foot, and I have frequently had my presence of mind upset by finding myself sitting down beside them. No accident happened, however, from their abundance, though there were many narrow escapes.

The horror with which the natives regard these venomous creatures is very great. I remember well on one occasion how I scattered a whole village, and my men besides, by coming amongst them holding by the neck a large green snake, eight feet long, which I had stunned and then picked up. It wriggled itself round my arm and body, though of course it could not bite, and when I appeared in this fashion, the people broke and fled in astonishment and fear. Before I could disengage myself I had of course to kill it.

On the 20th of June we emerged from this dreary waste, and entered a more undulating piece of ground covered with quartz pebbles. We crossed a delicious crystal stream, flowing between richly clad banks, with a sandy bed, the first clear running water we had seen since leaving England. The vegetation in its rich and varied luxuriance reminded us by its creepers and fine trees of the Usambara mountains, and the whole country seemed to have put on a holiday dress to receive us, after the filthy swamps and marshes of Uzaramo, and the deserts of the Rufiji valley. We passed through rich fields of ripened or ripening grain, with natives busily preparing the virgin soil for a second crop. As we passed and repassed the small stream which waters the surrounding plain, we drank deep and repeatedly. Passing through a perfect tunnel in a tropical forest with its grateful shade, we stepped into an open space winding in a labyrinthine manner among the trees, and dotted with houses which formed the delightful village of Behobeho.

Johnston's eagerly anticipated haven was thus reached, and hope and pleasure beamed on his face as we laid him down under the cool shade of a native hut.

Having entered a new country, and among different people, we may now profitably take a retrospective glance at the country we have passed through, draw isolated facts together which may have been lost sight of in the narrative, and present a condensed view of the whole.

A few words will suffice to delineate the topographical aspects of the country.

Near the sea we have the raised beaches already described, forming a comparatively level plain, rising to a height of not more than 200 feet. This plain presents few prominent surface features, the only irregularities being the small ridges and hollows formed by the insignificant streams which flow north to Dar-es-Salaam, east to the coast direct, or south to the river Rufiji. There is no scenery that the eye can rest on with pleasure, owing to the gentleness of the undulations and the absence of hills.

The landscape is made up of sandy ridges, covered with dense matted bush and stunted trees, monotonous and uniform in colour and form. The sides of the ridges are usually the cultivated parts, where the millet, the sweet potatoes, and ground-nuts are grown. The hollows are occupied with filthy marshes, where only spear-grass and huge sedges find a congenial soil among the rotting vegetation which fills both water and air with noxious and malarious gases.

About the villages a little more variety is noticeable. Within fifty miles of the sea many of the fruit-trees of the coast have been introduced, notably the cocoanut, forming groves round the villages, mixed with the Indian mango, the fragrant orange, and the curious jack-fruit.

In the wet season the country is entirely submerged, with the exception of the ridges, and becomes almost impassable. In the dry season again, owing to the sandy nature of the soil and the good drainage, the country becomes scorched up. The streams get dried, and everything presents a new aspect. Consequently the traveller can never be too cautious in the conclusions he may come to regarding the country.

One traveller may see it during the rain, and he has nothing to talk about but the swamps and marshes, the fetid mud, and the dirty water, till he has convinced every one that the country is of a painfully damp and humid nature. A second traveller may arrive four months later, and he has a different story to tell. He finds nothing but yellow shrivelled-up vegetation on the higher grounds, and tall impenetrable grass in the lower. The whole country bears the marks of an arid desert, and water is only found with difficulty. He then tells his lugubrious tale. He has had hardships for want of water! He has looked in vain for anything green! Not a sign of healthy vegetation has appeared!

It is thus with Uzaramo. To have its climate and character comprehended, it must be seen at different times of the year; in the same way as a plant to be properly described must be seen in the various stages of its growth.

Between Mkamba and Behobeho an even more unpromising country is found. It is almost perfectly level, devoid of watercourses, with not a ridge or hollow, and covered with open acacia forests. The grass is usually short and sparse, and nothing whatever relieves the landscape. Throughout the country traversed we found no sign of rocks, as it has been so recently raised from the sea that the streams in their denuding action have not as yet cut deep enough through the over-lying red sands which cover the lowland regions. But there can be little doubt, from an investigation of the country west of Behobeho, and from our researches further north in Usambara, that carboniferous sandstones lie underneath.

The Wazaramo, who inhabit the country between the rivers Rufiji in the south and the Kingani in the north, present no marked characteristic to separate them from the Waswahili of Zanzibar and the immediate vicinity of the coast. When rich enough they dress in a similar way. The greater scantiness of cloth arises solely from the greater difficulty of obtaining it cheaply. Their houses and customs are similar, and nearly all can speak the Swahili language. Their peaceful and unassuming character is amply shown by the fact that they seldom carry arms, unless they are on a journey. No attempt was made to purloin any of our articles, or to deceive us and throw obstacles in our way. On the contrary, they received us in their villages with hospitality. The verandahs of the huts were placed at the disposal of our men, and no attempt at extortion was made. The people showed great eagerness to enter into trading relations, and were most friendly in all their dealings.

The interesting fact, however, to be noticed in connexion with this, is the extraordinary change which seems to have taken place in the character and condition of the Wazaramo since Burton visited them in 1857, on his great journey of exploration to the Lake Regions. I think there can be little doubt that the description he gave of them was substantially true at the time, even after we have made every allowance for his condition as a "confirmed invalid," and for the jaundiced eyes with which he seems to have viewed everything pertaining to Africa and its people.

He speaks of well-stockaded villages; of apparent hospitality converted on the slightest excuse into spoliation and violence; of extortion by chiefs, which, if resisted, will be followed by flights of poisoned arrows. They are represented to be "the most dangerous tribe on the route to the Lake—ill-conditioned, noisy, boisterous, violent, and impracticable; caravans hurry through their lands, and hold themselves fortunate if a life is not lost, or if a few loads are not missing." If this was so, as, indeed, we have no reason to doubt, surely a mighty moral revolution must have taken place within these few years—a revolution which cannot but refute Burton's own estimate of the essential *character* of the East African tribes, among whom the Wazaramo seem to hold the lowest place, according to his description. Thus he speaks of their "apparent incapacity for improvement." "There is no rich nature, as in the New Zealander, to improve." They have "stopped short at the threshold of progress," &c., &c. And yet, in the face of these statements, we have the fact before us, that, in the space of twenty-three years, they have gone through a remarkable revolution of character. We no longer meet with stockaded villages, or extortions by chiefs; the people walk about without arms, turning their attention solely to trade and agriculture; caravans are received with genuine hospitality, and helped on their way. Is there no improvement in all this? If not, what can improvement mean?

It may be an interesting sociological problem to discover if possible the causes which have led to this change. There may be many such causes; but, of the more important, two may be mentioned—first, the influence produced by the consolidation of the Sultan's power along the coast; second, the increased desire to trade with the products of their own country.

When Burton passed through the country the influence of the Sultan in the government of the coast towns was little more than nominal. Each governor made the most of his position to enrich himself at the expense of the natives, and acted as the chief or sultan of the surrounding district. Instead of co-operating with the neighbouring governors, he looked upon them as his enemies, diverting so much trade to their own places, and seizing so much spoil from the natives, thus affecting his interests. The Wazaramo were naturally made the scapegoats in this state of things; plunder and murder were the order of the day. If they entered into a compact with one governor they were sure to be attacked by another. They lived in hourly fear of their lives.

In such a state of existence little wonder that they became treacherous and suspicious, built stockaded villages, and received strangers inhospitably. Too weak, and without union among themselves, to attack the nests of robbers which filled the coast towns, they naturally resorted to the trade routes, and revenged their wrongs, as far as lay in their power, on the caravans which passed through their country. Traders they looked upon as their enemies. Need we be surprised at the state of affairs Burton describes? The natives were not naturally what he found them, as he leads us to believe, but were made so by generations of brutal oppression and violence.

This state of matters soon changed. Ivory rapidly disappeared, and the slave-trade was stopped. These occurrences produced a wonderful revolution. The very existence of the towns seemed to depend on the traffic in ivory and slaves. But the former could only be got from great distances inland, and it became of no use to bring the latter in thousands to the coast when the export trade was stopped. This formed a great crisis in the history of the coast, and it seemed for some time as if ruin was their fate. But the traders had made the towns their homes, and besides they knew not where to go to better their condition.

Under these circumstances they were perforce compelled to turn their thoughts to more legitimate trade. There was a great demand for rubber and gum-copal. Over the whole of Uzaramo both were to be got in abundance. But how were they to be exploited? The traders with Arab blood in their veins, could not demean themselves by manual labour to obtain these articles, and the natives held possession of the ground. No other course being open to the traders, the Wazaramo were encouraged to come forth from their stockaded villages to gather the rubber and the gum-copal. The Sultan's influence had by this time become more marked in the government of the coast towns, and some form of justice had been established. The Wazaramo were encouraged to come into town for trading purposes. It was now seen to be for the mutual advantage of both to have peace and free communication with each other. Stockades became unnecessary, and gradually disappeared. The Mzaramo could go into the town or the trader into the country with perfect safety.

If the Anti-Slavery Society require an argument for continuing their good work, they have it here; for undoubtedly the stoppage of the slave traffic formed the turning-point in the great moral and social revolution which has taken place among the Wazaramo. We have no right as yet to come to rigid conclusions about the character of the negro, and what his capacity for improvement may be. Travellers who have made such sweeping denunciations of the negro have seen him as degraded from ages of exposure to the curse of slavery, ever fighting like a wild beast for his very existence—his hand against every man, and every man's against him.

I ask any one who knows anything about Africa to look over the land, and ask himself if there is not abundant proof of the improvability of the native. Compare many of the tribes with the Wanyamwesi, among whom the thin edge of the wedge of improvement has entered in the shape of a great desire to trade; or these again with the Wazaramo, among whom trade has still further advanced, and slave-trading and war have ceased, and over whom the peaceful shadow of a strong and improved government has fallen. May I not also point to the Wa-swahili as a further argument of the improvability of the negro? Their intelligence is beyond question, and the course of this narrative will show traits of honesty and faithfulness which would reflect credit on any people, whether civilized or uncivilized.

A few words with regard to the export trade of the country, and we have done with Uzaramo and its people. The only natural products of any value are rubber and gum-copal. The former is derived from two species of landolphia, called by the native mpilè and mbungo, of which the mpilè is considered the better. They occur as tangled woody creepers hanging from tree to tree in the richer parts of the forests. The rubber is extracted by cutting the bark, when the juice exudes like milk, thickening during the process. Then it is gradually taken off and wound round into the shape of a ball two inches in diameter. The rubber from Uzaramo is considered as good as any to be got either from any other part of Africa or from America, and would command as good a price but for the abominable practice of adulteration which the natives carry out. While making up the ball they introduce rubbish of all sorts—sand, hair, grass, or moss—with, of course, a nice coating of good material outside. This depreciates the value considerably, as the impurities are difficult to remove, and in consequence purchasers are chary of buying, except at very low prices, and thus, when an increased demand might have been expected, rather the reverse has occurred.

The trade in the valuable gum-copal of East Africa has also remained undeveloped, owing to other suicidal processes. Gum-copal, as most readers will be aware, is a fossil or semi-fossil gum resembling amber, found in the reddish sands which cover the low-lying country near the coast. The tree from which it exudes is practically extinct, though occasionally specimens are seen near the sea. The gum of living trees is of little value.

Curiously enough, the msandarusi, as the natives call the tree, seems never to have grown any distance from the coast, as no copal has ever yet been found in the heart of the continent, neither has the living tree. Livingstone, it is true, on several occasions speaks of an abundance of msandarusi-trees occurring to the south and east of Tanganyika, but certainly this has been a mistake. I made every inquiry possible among my men, and told them to be on the look-out for the tree, but without result.

Arabs, stationed in the very district referred to by Livingstone, declared they had never seen it. Chuma who had accompanied him bore similar testimony. Clearly some other tree had been mistaken for it.

Some of the varieties of gum-copal from East Africa are the very best that can be obtained, more especially the kinds used to coat paintings and give the last touch to coaches, &c. The unsystematic and arbitrary ways of the natives, however, keep the price up to such an extent as to drive every one from the market. The natives will only dig when and how they please. They have certain seasons when they exploit the copal. Sufficient only is dug out to buy all the cotton or beads they desire, but even then they are so independent that if they do not receive the exorbitant price they set on it, they are content to wait another season for a higher price.

The mode of digging is characteristic of the native. Proceeding to the forest, he looks about him for a place which appears to his eyes promising. He then takes a sharp-pointed stick, and begins digging a round hole, 8 inches in diameter, vertically downward as far as his arm will reach. If he be successful, he digs a few more holes near the same place, but if he is unsuccessful one or two times, he gives the place up and looks out for another, to resume work in the same manner. Till some more systematic and profitable way of working the ground is devised the trade in copal cannot be expected to flourish. I would suggest to some of our enterprising speculators who are ever devising schemes for opening up the trade of Central Africa, who talk about roads and railways into the interior for the transport of the mythical "unspeakable riches" to be found there, that they should moderate their wild notions and notice what lies at their feet. They would find it more profitable than conjuring up castles in the air. Uzaramo lies near the coast, and offers a field of enterprise of an exceedingly promising character.

Besides rubber and copal there is little else exported. Some grain finds its way to the Zanzibar market, and a hopeful trade is rising in sem-sem, or sesamum, which is exported to Marseilles, from which it reappears as the best olive oil.

CHAPTER V.

TO THE MOUNTAINS.

Behobeho, the village which we had now reached, well merits description, not only from the beauty of its surroundings, but also from the sorrowful event which marked our stay there.

Let the reader figure to himself a forest of the densest nature, formed of colossal trees, with deep green shady foliage, among which that prince of African trees the mparamusi or yellow-wood, with its silvery-grey trunk, rises prominently in stately grace. There are feathery acacias and mimosas, branching hyphene palms, and fan palms with their abnormally bulged trunks. A score of other species attract our attention, but their names were to me unknown. Fill up the intervening spaces between the trees with ivy-green shrubbery, until not a clear bit of ground is seen, and passage through the forest is rendered impossible. From tree to tree hang creepers of every description; slender leafy kinds, swaying gracefully in the breeze; giant forms thick as a man's thigh, gnarled and twisted, binding the tree-trunks as with bands of iron. The whole forms an impenetrable mass of vegetation, through which it is impossible even to see. Here and there, where a break occurs, the creepers may be perceived hanging snake-like from an overhanging branch, as if ready to strangle the unwary traveller, or forming light festooned bridges from tree to tree for gamboiling monkeys to cross. The hoarse cry of the hornbill, or the bark of baboons, are the only sounds which are heard from the forest, though, when darkness sets in, crickets with their fairy chirp, and the weird warning voice of the owl, or the croak of the frogs, help to break the stillness.

Such is the forest of Behobeho. In the heart of this dense mass of vegetation there is a winding open space, dotted with trees, labyrinthine in its character, here opening into wider areas, there leading into deceptive *culs-de-sac*. This space is dotted over with native huts, regardless of order, which form the village. The dwellings are of a different style from any we have as yet seen. The quadrangular huts of the Wazaramo are represented here by simple circular ones, with low walls, huge conical roofs, and broad overhanging eaves. Few of them are more than eight feet in diameter, and as there are neither chimneys nor windows, the smoke of the fires must escape by the low doorway. Among these huts you may observe the natives at their several occupations, attired in scanty loin cloths, and with undressed hair.

Passing through the tunnel which leads from the village to the outskirts of the forest, we get a delightful view, which might well remind us of our own Scottish land, in its character of hill and dale, clear sparkling stream, open glade studded with stately trees, and cultivated field. Let us take a commanding position, and look about us. Away towards the east extends an immense plain, apparently covered with luxuriant and noble forests; but we have already crossed it, and know differently. That light green tract is the burnt-up desert of the Rufiji Valley covered with acacia scrub. That darker patch we know to be simply dense bush, and it requires little imagination to make up the details hidden by distance. We are even tempted to strain our eyes to distinguish the sea, away in the far horizon. Turning south, the country appears to be less covered with forest. We are looking at right angles across the Rufiji Plain, and the natives tell us of lagoons, and swamps, and festering vegetation. Veering more to the west, a range of picturesque hills, running north and south, gives a pleasing relief from the monotonous tract we have been looking at. One hill or mountain specially arrests our attention by its rugged face and curious boldly-shaped outline. This range is doubtless the southern extension of the Marui Hills, which rise about thirty miles west of Dar-es-Salaam. Looking directly west, we observe through a gap in the hills in front of us, a magnificent series of giant mountains, which speak of cool breezes and a healthier atmosphere. These are the mountains which flank the great Central Plateau of Inner Africa.

Having thus reached Behobeho with such pleasant surroundings as these, new hope for our leader's recovery rose within us. Here surely were all the elements to please the eye, and by their charming novelty to infuse new vigour into mind and body! It soon, alas! became evident that we had come too late. We built a quadrangular hut for Mr. Johnston, as being more cool and commodious than the tent, and into this we removed him. He was, however, sinking fast, and little or no food passed his lips. To add to his agonies, a dreadful convulsive cough, the result of extreme exhaustion, rarely left him. We made but poor advisers. I myself had not the remotest acquaintance with illness of any kind, and could give no advice as to treatment, and I suppose I was but a rough nurse, though I did my best.

To employ some of my spare time during this detention, I made a short excursion to the prominent mountain already mentioned, as occurring S.S.W. of Behobeho. Crossing the small stream which winds beside the village, we traversed a broken piece of country, passed over a high ridge of sandstone which lay in front of the mountain, and reached the base of the latter, where a small stream occurs flowing south to the Ruaha. Here I for the first time got a shot at a herd of antelopes. I have no doubt I missed, though at the time I flattered myself that I saw one looking rather shaky on its legs.

I was very much struck by the symmetrical appearance of the mountain, which rejoiced in the name of Mkulima-hatambula. It rose like some magnificent cyclopean monument. It is quadrangular in shape, the upper half sitting on the lower as a base, with a flat terrace round the bottom of the former. The sides of both the upper and the lower parts rise almost perpendicularly, and give the whole such an artificial appearance as to suggest the idea of a huge monument.

I attempted to ascend to the top, but became so sick that I had to give it up and return.

On examination the peculiar shape was seen to arise from its geological structure, which was that of a series of lava beds intercalated between beds of fine chocolate-coloured sandstone in the lower part, and greyish-red coarse sandstone in the upper. The lava not being very decomposable, has resisted denuding influences longer than the sandstone, which has got worn away, till a broad, flat terrace of about half a mile lies between the base of the upper and the edge of the lower part. The sandstone which forms these hills belongs to the carboniferous system, and is found stretching along the coast from about the Equator to the Cape. The intercalated igneous rocks are volcanic, and contemporaneous, lying conformable to the sandstone.

I have taken the liberty to substitute the name

of Mount Johnston for the uncouth one of Mkulima-hatambula.

I at this time became subject to attacks of ague, which came on regularly at four o'clock in the afternoon.

On the night of the 22nd June the camp was thrown into an uproar by the arrival of men from Dar-es-Salaam, bringing our letters. It was a hard struggle for poor Johnston to get through his. To me there is something inexpressibly touching in the idea of a person situated as he was, trying with dazed eyes to read the many pleasant inquiries from friends at home, the hopes of a successful expedition, and that he was enjoying good health, and yet feeling that he was rapidly sinking into the grave, with his great work prematurely closed.

He now became frequently insensible, and gradually grew worse, until the 28th, when he finished his career. For the first time in my life I saw death, and I felt myself alone to take upon me the great responsibilities of leading what appeared to be a very forlorn hope.

One of the most promising of explorers who had ever set foot on African shores, Johnston has met his fate, and is numbered with the long list of geographical martyrs who have attempted to break through the barriers of disease and barbarism which make the interior almost impenetrable. It needs no words of mine to establish the fame of my late gallant leader, or to prove his unequalled qualification for the work laid out for him. These are well known to every one. But this I will say of him, that his whole soul was in his work, that not the slightest opportunity of adding to our exact knowledge of Africa was missed. Night and day he was ever on the alert, even when tortured by disease, and never satisfied except he himself saw everything done. Full of enthusiasm, and in every respect a scientific traveller, he would have led the Expedition in a clear, well-defined pathway. Without him the way seemed dark and uncertain indeed.

The position into which I was thus thrown was one of peculiar difficulty, and the question arose within me whether I should go forward or not. I was myself ill with fever. I was almost totally destitute of the special scientific knowledge of a geographical traveller; in fact, I knew little of anything that was most needful to know; and my age was but twenty-two. But though the question arose, it was soon disposed of. With my foot on the threshold of the unknown, I felt I must go forward, whatever might be my destiny. Was I not the countryman of Bruce, Park, Clapperton, Grant, Livingstone, and Cameron? Though the mantle of Mr. Johnston's knowledge could not descend upon me, yet Elijah-like he left behind him his enthusiasm for geographical research, and I resolved to carry out his designs as far as lay in my power.

It would not do to let the men imagine that there was any hesitation about my future movements, and

I stepped from the hut with my purpose distinctly defined. A basket coffin was at once constructed, and a space cleared in the dense forest. On the day following our leader's death we laid him in his last resting-place, where his grave is now green, as his memory will ever be. He lies at the foot of a large tree festooned with graceful creepers, under an arbour of dense evergreen bushes. His name and the date of his death are carved on the bark of the tree, and the chief of the village has undertaken to keep the place clear—a contract, I have since heard, he is faithfully carrying out.

Owing to the illness of Johnston, and our frequent stoppages to collect food, our progress had been exceedingly slow, so that it became necessary to lose no further time. The men were called together, and the new state of things explained to them. Boxes were overhauled, and letters written for the coast, and at last we were ready to resume our march.

There were two routes open to us,—one south-west through an uninhabited country on which no food could be got for six days, and the other a round-about road through Khutu, which, however, had the advantage of plenty of food. This latter was accordingly decided upon as the safer of the two, and on the 2nd of July we were once more pressing onward.

On getting out of my tent in the morning the prospect seemed rather discouraging. I was in such a condition that my brain reeled; my legs were so weak that I incontinently sat down, to prevent a fall. A few minutes in the open air, however, steadied me; and though I required some assistance to keep in a straight line when I started, yet I gradually gathered myself together, and got on wonderfully well.

Our route led us through the gap in the hill-range, to the west of Behobeho. We crossed the Grumby stream, and then an arid uncultivated ridge, reaching a pleasant valley, along which we marched, and having a range of hills connected with Mount Johnston running parallel to our route. A clear sparkling stream flowed beside us, and the scenery was pleasant, being diversified with forest belts, and open tracks of jungle. On our left lay the wooded hills, and away to the north and west rose the magnificent peaks and domes of the Duthumi Mountains. At midday we camped beside the stream of Vilanzi in a grove of bamboos; after crossing some beds of fine grey shale dipping W.S.W., coarse grey sandstones, and conglomerates, with sundry masses of volcanic rocks.

During the evening the camp was thrown into consternation by the cry of war on the path. "A ferocious slave-raiding tribe, called Mahenge, were on our route! Roads were shut up, and further progress impossible!" Rumour declared that we should be attacked without compunction, and that there was nothing left for us but to turn back.

Here was a pleasant bit of news! If we turned back where could we get another route? And yet if we went forward there seemed to be nothing open to us but fighting, as the warriors, in token of their blood-thirstiness, were painted in all the dreadful colours which a savage imagination could suggest.

On cross-examination, however, I could gain no definite intelligence as to the exact whereabouts of the Mahenge, or their special object in making war. I therefore determined to go forward till I got more reliable information. It was necessary, meantime, to speak to the men, as a stampede might take place at any moment; indeed, as it was, I quite expected a number of desertions in the face of these rumours.

I therefore explained to Chuma that as yet there was nothing to fear, but that the utmost caution in our movements must be maintained. Every one must be on the alert, and shooting or noise of any kind must cease. I gently hinted that any one caught running away, or raising a panic, would run the risk of being shot without any parley. These facts I directed him to explain to the men.

As the evening shade was falling, and the ruddy glare of the fires lit up the scene, the drums were beat, and an eager throng of men gathered round to hear what was to be done. Chuma took up a prominent position on some bales, and commenced his harangue. He had not got through three sentences till I felt that I was listening to an orator, as he threw his whole soul into the work. He swayed that gang of rude savages as no refined and polished speaker could have done, and roused their enthusiasm in such a way that I felt there was little reason to fear desertion that night. Makatubu followed with almost as effective a speech. Then one of the Kiringosis spoke up for the men, and stimulated their courage still further, as he swore that they would stick to me through thick and thin—a sentiment re-echoed all round by a chorus of "Eh Wallah!"

On the following day we continued our progress, pushing slowly through rather trying tracts of tall jungle grass and tangled belts of rich forest, which circumscribed our view for hours together to a few yards. It was clear that as we approached the edge of the Plateau the vegetation was becoming richer; a fact manifestly attributable to the frequency of the rains and the numbers of the streams.

At midday, after crossing a stream called Mtambo, flowing north to the river Mgeta, we reached the village of Mua. This village, like Behobeho, is impregnable with its forest-bound walls, through which an enemy would require days to cut a way. It is reached by a narrow pathway, guarded by two strong gates. We found the villagers in great terror regarding the Mahenge, but entirely ignorant as to their precise movements. They might descend on us or on a village without a moment's warning. We were, however, rather cheered by a piece of news we got here. It appeared that the Mahenge were not directly attacking the Wakhutu, their operations being ostensibly directed against an insignificant tribe called the Walugulu, inhabiting the high mountains some twenty miles north of Mua.

I was compelled to stay here a day to recruit, as the fever which had been hanging about me since I left Behobeho had become worse.

As I felt somewhat improved by my rest our journey was resumed. We were still marching parallel to the hills on our left, and occasionally we got peeps of the distant mountains to the north. Marching quietly along in single file through a dense piece of forest, with not a sound to break the stillness, and every one on the alert, we were suddenly startled by a stifled sound proceeding from the forest. An immediate halt took place, and I could observe a sort of nervous shock pass along the line of porters, till at last an articulate whisper of "Mahenge! Mahenge!" reached my ears. The foremost men were already hurrying back. Loads dropped from paralyzed shoulders, and a panic ensued. Fortunately I was in the rear. So, seizing my gun, I raised it and threatened to shoot the first man who attempted to pass me. This new danger fairly dumbfoundered them. Some, in spite of torn clothes and bodies, attempted to escape sideways, but nature favoured us too well. At last, with the invaluable aid of Chuma and some of the more courageous of the men, the caravan were somewhat quieted, the loads gathered together, and the men around them.

With a considerable feeling of relief we found that we had not been observed. This was a fortunate thing for us, as the Mahenge would certainly have swooped down and completed our flight if they had noticed our attempt to retreat. In such a case few of the porters would have stopped till the coast was reached, with the story of a mighty battle, in which I would of course be killed. However, we had now time to breathe, and think what we should do next. A council of war was accordingly held, and we came to the conclusion that this state of things must not last, or ruin would come upon the expedition. It was clear we must come to some understanding with the Mahenge, and learn whether they were to be friends or enemies. I determined to try the effect of my white skin, and to show complete confidence in the savages. Selecting Chuma, and a porter who could speak the language, we went forward unarmed, leaving instructions for the men to be ready for any emergency; and to tell the truth, they seemed wonderfully courageous after their first panic.

In a few minutes we got clear of the forest, and a strange and unusual scene met my eye. Through the long grass tramped in single file a long line of warriors, dressed in the most wonderful feather head-dresses, with a few wild-cat skins on their shoulders, but otherwise entirely nude. Their faces were painted in the most hideous manner, and certainly they did look like typical savage warriors. Their arms consisted of a stabbing-spear, two or three assegais, a club, and an oval or elliptically-shaped shield of bullock's hide, which they were carrying over their heads to protect them from the wet grass. It was really a sight worth seeing.

On emerging from the forest we raised a shout to attract attention. In a moment they halted in alarm. Down came their shields; spears were seized, and everything got ready for a fight. Some dropped behind bushes. It was a critical moment, but we could not now retreat. I stepped into view, and my appearance took effect on them like a shock of electricity. A buzz of surprise arose. I lifted my hands to show I had no weapon, while my interpreter shouted out "Mzungu! Mzungu!" and declared that we were friends. At once suspicion was disarmed. They gathered round us with shouts, and surveyed me as a curious animal about which they had heard strange stories.

After we had obtained silence, we proceeded to explain who we were, where we were going, and that our intention was to make friends with them. They at once assured me we were quite safe. They were not fighting the white man. They had been away fighting the Walugulu, but as their chief's son had become ill, they were escorting him home. They further declared that we should be made most welcome in Mahenge, when we arrived there. This was all very satisfactory, and relieved my mind very much.

We sealed our friendship by "making brothers" with the chief's son. This ceremony is done by tasting each other's blood, and vowing eternal friendship. Of course I did not take a personal part in this savage ceremony. Chuma acted as my proxy. The caravan was then called up, and a salutary thrashing was administered to a few who had run away on the first alarm. Thereafter we fraternized on the best of terms, and proceeded on our way.

The Mahenge, however, were so wild and savage-looking that I was in momentary fear of a ruse, and therefore sent a word of caution among my men. The conduct of the warriors was indeed very suspicious-looking. They went rushing backwards and forwards, mixing here and there, and seemingly getting ready to strike a blow. Our fears luckily proved erroneous. We camped at Mwigonga, and on the following day we parted company with the Mahenge, though two of them were left as guides. On the 7th of July we reached Kilengwe, one of Burton's camps on his celebrated march to Tanganyika.

I had now become reduced to such a state that I could not stand. On moving out of my tent to try to start, I fell twice, and had to crawl ignominiously back into my tent. It became absolutely imperative that I should rest, and as I had to depend upon my compass bearings to protract my route, I did not care to be carried.

It may not be out of place at this point to say a few words about the Wakhutu, and contrast their condition at present with what it was when Burton passed through. That man of learning, though he seems to have the vocabulary at his finger ends, is hardly able to get strong enough language to describe the miserable condition of the natives. They live, he says, in huts, "composed of a few short sticks, tied together at the top, and loosely covered with a few armfuls of holcus stalks." "Their largest villages consist of three or four hovels." When I passed through, their material condition had certainly improved. I found their huts large and commodious. The three or four families living in miserable hovels have increased to large villages of two or three hundred houses; and they have emerged from their hiding-places amongst marshes and jungles, to live in impregnable though charming tracts of forest, where nature has exhausted her resources in producing everything that could make the place pleasing to the eye. They live in the richest piece of country to be found in East Africa, where perennial showers and numerous streams ensure continued production of food with a minimum of trouble.

But though they have thus improved in material comforts, they have made no social advance, no step towards civilization. They are now as Burton found them, one of the most degraded tribes to be found in Africa. With their black sooty skins and miserable withered bodies, they present such a picture of apathy and wretchedness, in the midst of so much natural riches, as appears paradoxical. Nothing disgusted me more than to see them gather round me in crowds, sitting doubled up like so many ancient Egyptian statues, watching me with an idiotic lack-lustre gaze, and looking as if they were so many slave-gangs resting on their way to the coast, with all hope of life and liberty flogged out of them. Nothing I could show them elicited the least mark of surprise; everything seemed alike to them; and without doubt they were beneath the monkeys in the extent of their curiosity.

The dress they wear does not differ much from that of the Wazaramo, except in its greater scantiness. Those who can afford it have the simple loin-cloth, but many are reduced to the possession of a small flap of cotton, drawn over a waistcord before and behind. A few, however, have nothing better than a grass kilt.

It is not difficult to account for the low position which the Wakhutu hold. Being an exceedingly timid race, they have been continually made war upon, but have never shown any aggressiveness in return. The Wazaramo, the Waswahili, and the Mahenge have made the Wakhutu the victims of their slave-hunts for ages past; and not having the power or the spirit to retaliate, they have never risen above the lowest grade of barbarism. They still remain in the stage of savage life in which the struggle for existence absorbs the whole faculties.

There are two great steps which every tribe must take before it gets on the highway of improvement. There must be a social aggregation for mutual protection and safety, and there must be a differentiation of the social aggregate, so that by an exchange of services they may become mutually useful to each other. The idea of chieftainship is at the root of the former; trade secures the latter. Now among the Wakhutu there is no common union for protection. A few families perhaps collect together and settle down in a village, and nominally recognize a headman among them; but with individuals of other villages they have no common action or mutual dependence. Hence, when an enemy appears requiring the whole strength of the tribe to cope with him, there are no means of securing unity or common action. Each village has to depend on itself, and there is nothing for it but to flee for their lives into the jungles and forest.

African tribe life teems with proofs of the fact that with the progress of organization and social aggregation there is a corresponding advance in social improvement. There first arises a controlling chief, who acts as the "thin end of the wedge" in the mass of barbarism. He brings village after village under his influence, till his country becomes too large for personal government. Sub-chiefs are chosen, classes are formed, and as the tribe grows stronger, safety for life and property becomes assured. New wants are raised, and thus step after step is gained in the ladder of progress.

But, as has already been remarked, along with absence of social aggregation and governmental organization, there is in the case of the Wakhutu an utter absence of differentiation among the individual units. There is no mutual dependence or exchange of services. Each man builds his own hut, makes his own weapons, and digs his own fields. He has no wants which he cannot supply himself, and hence with no necessity for trade, none exists; and thus shut up in himself, and cut off from the outer world, he retains his primitive barbarism, and no one is interested in his individual existence, except as a means of satisfying rapacity.

These are the principal reasons why the Wakhutu remain in the same degraded state in which Burton found them, and doubtless the slave-trade was the main cause of it. Since that ceased a day of better things seems approaching. For, though still troubled by the ferocious Mahenge, they are now left alone by the Wazaramo and Arab traders, and the effect we see in the increase of their material comforts.

Unfortunately for the Wakhutu, while, as we have seen, nature favours them in one sense, it fights against them in another. Their country lies among the hills which stretch along the base of the high coast mountain ranges. These, attracting the fleeting clouds, produce continual showers throughout the year. The earth thus drenched, while it yields abundance of food, becomes one great festering fever spot, emitting foul gases continually from the rotting vegetation, till the atmosphere is charged with malaria, and hangs like a pall over the land. The weather is ever varied by cold breezes from the mountains, alternating with damp, sickly, fever-laden winds from the east. These of themselves are sufficient to explain the physical debility and mental prostration of the natives; and indeed to me the greatest wonder is, that with man and nature conspiring against them, they have not been swept off the earth entirely.

However, with greater safety to life and property a new era of hope is, as I have said, dawning on them; for even the Mahenge, finding slave-hunting not so profitable as formerly, are beginning to be more peaceable and friendly. The Wakhutu will soon find out the advantages of trade. Their forests abound with rubber-trees, and many of the products of their fields will find a ready market at the coast; more especially tobacco, which grows in great abundance, and sesamum, from which a capital oil is expressed. The tobacco of Ukhutu has already a wide-spread reputation in the interior and at the coast; and small trading caravans of Wanyanwesi not unfrequently come all the way from Unyanyambe for the "soothing weed," which they barter to the Wagogo and other tribes, the nature of whose country does not permit them to grow it themselves.

Greatly improved by three days' rest at Kilengwe, though still very weak, I ordered the men again to shoulder loads. An hour's splashing up the muddy waters of a stream called the Msegwe was not a very encouraging start to a person half dead with fever; but it had to be done, as there was no other way, except by cutting a passage for ourselves through the tangled forest and jungle, a proceeding which would have entailed too much labour. So I clutched desperately at the shirt-tails of Uledi, my servant, and floundered along. These well-used appendages, however, unfortunately gave way, and I had to get the assistance of another man for decency's sake.

Nothing more charming than the vegetation along the high banks of the Msegwe could be conceived. We appeared to be in a vegetable tunnel, with only sufficient light to show off everything. The thickly-foliaged trees on either side spread over the stream till their branches interlaced. The mparamusi, with its stately bole, shot up from the others, as if seeking the free air of heaven. The hyphene and ukindu palms added variety of shape and grace; while the creepers, ever so profuse in tropical forests, filled up interstices, tenderly decking the aged and crabbed trees with pleasant leafage, till the ugliest and most rotten trees looked beautiful. These creepers wound round colossal trunks and crept up among high branches, from which they would hang in straight or curved lines, interlacing with each other, and swaying with every breeze. If the stream had been but a good pathway, what a delight it would have been to wander along this fairy arbour, to lie and note its beauties, to watch the gambolling monkeys, and trace the fluttering course of the butterfly, or the more rapid flight of the dragon-fly! But I was weak with fever, and splashing deep in mud and water; so, of necessity, I must not linger.

After leaving the Msegwe, and having rested for an hour, we resumed our march over a piece of broken, undulating ground. Up to this time we had not had the pleasure of a walk over country of such a nature, and it gave some zest to our movements. Basalt appeared at the surface in a very decomposed state, and denudation had wrought havoc in the mouldering rock, cutting it up into deep, narrow glens, filled with a pleasant flora, consisting of ferns, shrubs, and trees, while on the tops of the ridges everything green was shrivelled up, even to the trees, under the fierce sun.

After a long march through a country of sharp ridges and narrow glens, we reached the stream of Viladzi, and camped in a grove of bamboos in a deep hollow, with the rippling waters at our feet. The sun had set when we stopped. The night was very dark, and the scene around our camp was of a nature so peculiarly enchanting that I lay outside watching it. The ruddy camp-fires light up the hollow with weird effect among the skeleton-like bamboos, throwing back the darkness to accumulate in impenetrable gloom around. Animated groups of laughing story-tellers crouch round the fires, awaiting the cooking of their supper. Some are moving about like ghosts in their white shirts; while others twang away for hours on rude stringed instruments, accompanying somewhat monotonous, though not unpleasing, songs, doubtless to their dark-skinned mistresses far away in the charming isle of the sea, from whom rude fate had separated them to wander through strange lands. Everything speaks of ease and comfort after the day's hard labour. Gradually, as night moves on and the stars traverse the heavens, the porters arrange themselves for sleep, wrapping their long sheets of cotton around their bodies, and stretching themselves out on their grass couches. The story-teller falls asleep, and the loud laugh is no longer heard. Still some love-sick porter may twang his instrument, but he sings with more subdued voice. Other sounds begin to assert themselves. We can now hear the evening breeze creaking through the bamboo forest, while the gentle, murmuring stream sings a soothing lullaby, and the multitudinous, ever-changing, fairy music of the cicadæ pervades the solitude.

Fortunately nothing the worse of our long march

to Viladzi, we recommenced our march to the chief town of Khutu. As on the preceding day, we had an hour's tramp along the stream, which flows between high mud banks covered with dense vegetation. Here and there we observed good sections of the rocks, showing the intrusion of basalt among the sandstone beds, which were much tilted and broken. Diverging from the stream, we crossed a broken piece of country, through a pass among low hills which formed a water-parting between the Kingani and Ruaha—the Viladzi on the one hand, and the Engomma on the other, representing each of these river systems.

Leaving the pass, we crossed the Engomma, and after traversing a very wide area of rich, well-cultivated fields of various cereals and vegetables, we entered the dense forest in which Mgunda, like other Ukhutu villages, is placed.

When Burton passed Mgunda it was a place of no importance, and was known as Mbwiga. Zungomero was at that time the only place of any note or size. Since then matters have much altered. The latter place is no longer the principal station in a trade route, as at that time it was. Even its name has disappeared; for it is now known as Kisake. Mbwiga, on the other hand, has, by an infusion of new and more energetic blood, not only become a place of some importance, but has come to be recognized as the chief town. It was taken possession of many years ago by a colony of Wanyamwesi, under a chief called Mgunda, who has given his name to the village. Wherever Wanyamwesi settle, they are sure to become an important factor, from their superior intelligence, mechanical skill, and love of trade.

There was no difficulty in recognizing a different tribe on entering Mgunda. In their energy and curiosity they presented a marked contrast to the Wakhutu. Their houses were better, as well as their own personal appearance. They dress their hair, which the Wakhutu do not do. This, however, only refers to the men, as the women leave their hair in its native furziness. Curiously enough, only the left side of the head is dressed, and there it, is shaved into all sorts of curious designs, in the style of an ornamental garden—circles and spots, squares, spirals, and triangles, diversifying the cranium in the strangest fashion.

Mgunda is situated at the head of a beautiful valley, with the picturesque peaks of the Rufuta mountains to the north, from which two spurs of low densely-wooded hills trend south, forming the valley of the Msendasi stream. Of these spurs, the one to the west forms the Mabruki hills, rising from the valley to a height of 300 feet. The valley is about thirty miles long by six to eight miles broad. It is extremely fertile, and watered by numerous streams. The only part now cultivated or inhabited is the extreme northern end, the rest having been laid waste by the Mahenge, who have thus transformed a perfect garden into a desert. So completely have these savages done their work, that, where once were thriving villages, not even a pathway exists through the jungle grass.

The day after our arrival the chief visited us, bringing a present of two oxen and two baskets of rice. The oxen were the first I had seen since leaving the coast, and it appeared that he had only four altogether. Mgunda is rather old, but energetic and intelligent. In conversation I learned many curious particulars about the surrounding tribes.

Among the hills on the left of our route from Behobeho to Mgunda live a small tribe of people called the Wamahala. He described them as few in number, living principally on honey, and fish brought from the Ruaha. They poison their arrows, and their language is said to be different from that of Ukhutu.

In the high mountains to the north-east live the Walugulu. Although dwelling in extremely high altitudes, it is said that they wear only banana leaves as a dress. They use spears chiefly in fighting, and owing to the rugged nature of the country they are difficult to reach. These are the people upon whom the Mahenge were making war. Mgunda informed me that he was also intending to join the Mahenge against them. On my asking for what reason, he coolly informed me that it was to make friends with them!

Away to the west and south-west live a third tribe, called the Wangwila, among huge mountains with strange unapproachable valleys, into which the people retire during war and cannot be followed.

I now learned that another stoppage was necessary, to collect food for a six days' jungle march to Mahenge, the intervening country having been depopulated by the warriors of the latter place. I thought three days would have been sufficient for our purpose; but owing to the rains and dilatory conduct of the men, it became five before we once more got *en route.* Although we were now in the height of the dry season, the rain came down in torrents.

The chief was building a fine new house, and I took occasion to visit it. In form it was like the usual Wakhutu huts, namely, a circle with a huge conical roof, which extended over the walls in broad eaves. The hut was divided into apartments by an inner circle, forming in the centre the chief's room and audience-chamber. From this doors led into sections of the space between the inner and outer walls, which were used as store-rooms, and also occupied by his wives.

The weight of the huge roof rests chiefly upon a central pole, from the top of which the rafters radiate to the walls. The walls are formed of upright poles, four inches thick and a few inches apart, bound together by creepers, and then thickly coated with mud. Of course such a thing as a window, or ventilation of any kind, is quite unknown, the low doorway not only acting as a means of going out and in, but ventilating and lighting at the same time.

The hut was receiving the last finishing touch of mud when I visited it; thirty people being employed for this purpose running backwards and forwards with single handfuls of the plaster, and singing lustily all the time. In the evening, when they had finished for the night, they came and surrounded my tent, dancing round in a circle to the exquisite (?) music produced by notched bamboos rubbed vigorously with a stick—a most exasperating whirring; and singing all the time "Our labour now is ended, and we must have pleasure." As I did not relish their modes of pleasure, I was glad to give them two yards of cotton to get rid of them.

We had here some capital fun in testing the shooting powers of the men. About one half of them had never fired a gun in their life. Some were so frightened that I could not get them to shoot; others fairly trembled with excitement till they could not hold their guns. They shut the wrong eye, or both eyes. One or two had been accustomed to shooting from the left instead of the right shoulder; the greater number of them exhibited ludicrous fear of being impaled by the ramrods in the stock of the carbines, and could not be persuaded to shoot till they were taken out. However, after some amusing hours we finished without an accident.

On the 18th of July we once more got in motion. Our route from the Rufiji to Mgunda had been generally W.N.W.; we now turned S.S.E., along the valley of the Msendasi. As has already been noticed, our way lay through a pathless jungle and forest. To get along at any ordinary rate it became necessary to organize a company of pioneers to cut down the huge grass, the tangled creepers, bound brushwood, and other obstructions. The headman and a number of the Kiringosis were impressed into this service. With axes, bill-hooks, and knives they worked with energy, clearing a way with surprising speed, so that we were able to move almost at our usual rate, though the nasty stubble and sharp ends of creepers made painful wounds in the feet and legs of the men, besides reducing their flimsy clothes to tatters.

I was forcibly struck during these jungle marches with the very marked absence of animal life of whatever class. This indeed had been observable over the entire country we had traversed, but hitherto we had been generally in populated parts, from which at least the larger game had been driven away. But now we were in a fertile tract of country almost never disturbed by the sight of man, and where game might roam unmolested; yet rarely did anything animate meet our eyes. True, indeed, we saw frequently tracks of elephant, buffalo, antelope, and quagga. Encouraged by such evidence we often roamed the jungle, but all in vain; they seemed to have deserted the country *en masse*. Men were sent out in all directions, but with the same result; no game was to be seen. It was the same with all other classes. In the forests we expected to see numerous monkeys, squirrels, and similar tree-loving creatures, but we peered among the branches without result. No birds flitted from tree to tree, or filled the air with their song.

In few parts of the world, with everything apparently favourable for the rearing of abundant animal life, is there such a marked absence of it without an assignable cause. Of course I can only speak for the time of the year in which I passed through it. The rainy season had just closed, and doubtless the grass was too rank and luxuriant for the larger game, which had probably migrated to more favourable grazing-grounds, as they are often in the habit of doing; and the fruits on which the birds, monkeys, and other small animals depended were scarcely ripe, and they also may merely have shifted their ground till a more favourable season came round. Taking into account my subsequent experience, however, it still seems to me a most astonishing and unaccountable phenomenon that there is such a marked scarcity of animal life in East Central Africa as seen by a traveller passing through the country. Of course it is simply as a passing traveller that I speak. Doubtless a naturalist settling down in some part of the country for the express purpose of studying the subject, might have a different tale to tell. Nature, we know, usually reveals her secrets to the select few who love and patiently search for them.

On our fourth day's march from Mgunda, through a varied country of hill and dale, our eyes were gladdened by the sight of the river Ruaha,—the first time it had ever been seen in its lower portion by any European traveller. The upper sources of it have been of course explored, from the north end of Lake Nyassa to Ugogo, by Captain Elton, in the journey which ended in his death.

At first sight it looked promising, but a more careful examination revealed a state of matters which at once put it out of the category of navigable rivers. The point where we reached it was a sharp curve. As far as we could judge, the stream was from 80 to 100 yards broad, flowing with a strong current. The deepest part was along the north side. But there it attained a depth of little over eight feet, and it flowed among nasty rocks, which would make boating extremely dangerous. The greater part of the river was less than eight feet in depth. But even more serious defects than these were observable. About half a mile above our crossing-place rocky rapids were seen. At about the same distance down the stream a barrier of sandstone stretches almost from bank to bank, leaving only some twenty yards for the passage of a swift current. A quarter of a mile further down a still worse barrier occurs, leaving the river to escape by narrow channels through which no boat could possibly pass. From all I could learn, it would appear that the entire river is blocked by such obstructions.

On arriving at the banks of the river we at once set about preparing for the passage. In three minutes Admiral MacDonald's little collapsable boat was ready for the water. Amidst great enthusiasm it was launched, and named the "Agnes," after my mother. Two men well acquainted with paddling were selected to work her, and in the experimental trip they started empty. Not at all accustomed to such a form of boat, they were not well off from the shore till they lost command of their craft, and every moment we expected to see a capsize, in their frantic efforts to keep her head up stream in pushing across. The prospect looked discouraging, and we had melancholy forebodings of loss of bales, if not cases of drowning, before we were all safely on the other side. They got back safely; but the time consumed led us to look forward to two, if not three days' hard work.

To assist the "Agnes," and facilitate our movements, we made several unavailing, though energetic, attempts to carry a rope across the river. We were thus compelled to rely entirely upon the boat. Two or three more trips put the men into better working order, and our hopes rose each time, as the boatmen acquired more experience in managing her, till we began to admire her powers. In the afternoon a rotten bark canoe came to our assistance; but the greater part of the work had to be performed by the "Agnes," which, between ten a.m. and six p.m., conveyed across 100 men and 120 bales, without damage or loss to a single article. We had a considerable amount of fun in conveying the men over. As they could nearly all swim, the "Agnes" was packed far beyond her floating powers, and only kept above water sufficiently long to take them over the most dangerous part. The water then getting in, she would begin to fill and roll, till she sank beneath the waters, among a group of spluttering men, a sight which was always greeted with screams of delight by the onlookers. Our jollity and excitement reached its climax in getting over our sole surviving donkey.

We had now entered Mahenge, the Ruaha forming the boundary-line between it and Ukhutu. But there were still two days of jungle and forest before reaching the inhabited portion of the country. The physical features of the ground afford little worthy of notice. On our right, at a distance of a few miles, rose a magnificently-peaked range of mountains, from 4000 to 6000 feet in height. Behind us were the low Mabruki hills, we had just left; while to the south and east appeared an apparently indefinite extent of jungle and forest.

On our way to Mahenge we had heard strange

stories about the extraordinary abundance of elephants in that country. They were said to be looked upon with such reverence that they were never molested, and had become so tame in consequence that the sight of a human being did not disturb them in the least. It appears that formerly there had been some grounds for such a story. A certain chief of Mahenge had had such a regard for these noble animals, that no one was permitted to disturb them. However, much to our chagrin, we found that even here, during the last generation, evil days had befallen the elephant after their patron died, and now not one is to be seen over the wide jungles which surround Mahenge.

On our second day's march from the Ruaha we entered the Mahenge village of Joto. The people, divested of their war-paint and feathers, presented much the same general appearance as the natives we had passed. Physically and mentally they were certainly far superior to the Wakhutu. Their curiosity is boundless. They watch every movement, and greet everything that tickles their fancy with roars of laughter, dancing about to let off the effervescence of their feelings at the sight of the strange doings of the white man. They are full of cunning, with a lie ever at the tip of the tongue.

In the afternoon the chief of the place, who ranks second in power, according to our informants, gave us a gentle hint that we had other people than the Wakhutu to deal with, by sending a message for four yards of fine coloured cotton, as he was naked, and could not, for decency's sake, come and see me till he was properly dressed. This is the common mode of begging among African chiefs, as our subsequent experience showed.

Next day, after much trouble about the carriage of sick men, we trudged for six hours through a waterless jungle to Pangalala, the chief of which was also described by our guide as the second chief in Mahenge. After the customary exchange of presents, the chief told me to stay next day and give his people time to see me. Not being inclined to be lionized, I promptly declined to do so; but what the chief could not compel me to do, the porters did.

In the evening they all gathered round me, evidently with something of importance on their minds. The chief Kiringosi at last stepped forth, and declared that they were not able to march to-morrow, and that they required a day's rest. They were not accustomed to the long and continued marches which they had had from Mgunda, and were all tired and knocked up in consequence. Some were half dead with rheumatism from the frequent deep swamps they had to wade. Others were so ill about the chest and back that they could hardly move. When he had finished his case, I told them that I wanted to go on only another day, when we would reach the big chief's town, and I pointed out the frequent long rests they had had at various places. This had no effect. I next asked them to continue their march as a favour to me. My appeal met with the same result. I tried threats, but they only laughed. They were unanimous, and the coast could be easily reached by a well-known caravan route to Kilwa! They saw they had me quite in their power, and as I felt it was useless to struggle against a united band, I was *nolens volens* compelled to give in.

In disgust I went off early to bed, and was just dozing off in the darkness, when I was awakened by a tumultuous uproar. Drums were being beaten, and horns blown. The zomiri screeched out its ear-piercing note. Singing and clapping of hands made up such a wild medley as is rarely heard. Going out to see what was the meaning of this strange business, I stood astounded at the sight which presented itself to my eyes. The whole caravan was out for a dance, and the broken-down looking band of men I had seen two hours previously, seemed now inspired with exuberant animation and energy. The rheumatic, the stiff-backed, and sore-chested, all seemed to have laid aside their ailments as an easily divested garment. Each one had acquired the agility of a dozen ballet dancers, and stamped and wriggled about like a madman. I returned sorrowfully to my tent, but slept no more that night. Three days later, after traversing the usual jungle and swamp, which even at this advanced time of the year covered the flat plains of Mahenge, we reached the well cultivated outskirts of Mkomokero, the chief town of Mahenge.

Passing a considerable village occupied by the wives of Komokero the chief, we commenced announcing our approach in the manner so dear to the heart of the African. Volley after volley thundered from the guns of the men. As they had nothing but ball cartridges, I was in momentary fear of an accident, but fortunately none occurred.

To mark their sense of the importance of the white man's visit, such of the natives as had guns honoured us by salvos, feigning at the same time an attack on our caravan. Now they would dart out unexpectedly from a hiding-place, and fire off their guns in a very alarming fashion. Then, seizing their spears, they would make a sudden rush, as if to finish off some wounded enemy. Retreating at a run, they would recommence the same tactics, keeping up a running fire of shouts and war-cries.

As we had arrived very early in the morning, and as it would be indecently hasty to ask to see the great chief on the first day of our arrival, I determined to occupy the time by a visit to the Uranga, a river which though long heard of, had never yet been seen by any white man. A walk of two miles through rich fields brought us to a wide stretch of back-waters, lagoons, and swamps, which border the Uranga, and which render approach to it a matter of difficulty. These back-waters teemed with fish, and formed the favourite haunt of innumerable flocks of all sorts of aquatic birds.

To reach the river-side we required to take a canoe. This proved to be most enjoyable. We lay lazily and paddled through placid ponds with swimming-birds on all sides, now pushing by a narrow channel through sedges, disturbing the crocodiles in their noonday siesta, anon causing the hippopotamus to sink out of sight with indignant snort. At last the river was reached, but the canoe-men would not venture on it, as the hippos were described as exceedingly dangerous. This was disappointing, but they would not be persuaded.

At the point where we reached the river it is divided by an island, so that its actual breadth could not be determined, and the huge bordering sedges cut off any possible view of it either up or down.

The part seen is about fifty yards broad, with a slow current and very deep, no bottom being found with a very long pole. The fishermen said that the other branch was very much broader, and that so far as they knew it was all alike on to the Rufiji. I think there can be little doubt that it is navigable for the largest river-boats as far at least as Mkomokero from its junction with the Ruaha. But beyond that I could get no information, except that it flowed from the S.W. through the country of Ganga. In that direction the people were said to be very few, and about those that did exist nobody seemed able to give information as there were no roads to make the country accessible. I had at one time thought of exploring the Uranga, but in the face of such a discouraging report I thought it advisable to give up the idea.

Next day a great waste of gunpowder heralded the approach of Komokero. First appeared the shooters, followed by four goats and a sheep led by men. Next came about thirty women, carrying as many baskets of rice on their heads. Then the chief himself, accompanied by two brothers, advanced with a huge mob at his heels. Komokero is moderate in size, has a bushy beard and moustache, and looks half an Arab. He is far from dignified in appearance, and there is a sinister cunning twinkle in his eyes. In dress he is little better off than his subjects, from whom he is only distinguished by an overpowering odour of rancid castor oil, so that in a crowd he is discovered faster by the nose than by the eyes.

In conversation all necessary forms had to be observed on such an important and memorable occasion. And this was the manner of it. First I made a speech detailing our objects in coming to and passing through Mahenge, telling of the advantages accruing from an acquaintance and friendship with the white man, and finally asking for a guide. I spoke in English to Chuma. He translated it into Ki-swahili to our interpreter, who recited it in the language of the Mahenge to the chief's son, my brother, according to the rite we had passed through in Ukhutu. The round was finished by the chief graciously listening to the words of his son, to whom he gave the reply, and thus in inverse order it got back to me a quarter of an hour after I had spoken.

The reply was far from satisfying, and somewhat alarmed me. Guides would certainly be supplied to me; but as a real white man had never before visited Mahenge I must stay a few days to give his subjects an opportunity of seeing me and studying my appearance and customs. To this I entered an energetic protest arguing my numerous engagements in the direction of Lake Nyassa. The travelling season was considerably advanced, and if I was detained in such an unlooked-for manner, I would be caught by the rain! It was all of no use. As a pioneer of civilization, and a living example of the blessings of ample clothing and frequent washing, I was constrained to bow to the inevitable and stay.

So a royal proclamation was sent over the country making it known in African fashion, that the chief, ever mindful of his loving subjects, had, regardless of expense, secured a real white man, and that all who desired to see this great curiosity must come at once as he could only be detained a few days. In response to this invitation the people flocked to the exhibition in crowds. They issued, miserable and sooty from the swamps and marshes to the east.

They flocked down in wild array from the high mountains to the west. The fishermen from the rivers Uranga and Ruaha sent their quota till Mkomokero was filled with visitors.

In high dudgeon I retired into my tent, and kept myself as free from observation as possible. But in that way I could do little, as it was so frightfully hot that the tent door could not be shut without suffocation. I at once became all the rage, and it would have quite delighted any philanthropist to see the way in which they studied my every movement. Even the mysteries of the toilet could not be veiled from their curious eyes, a fact which caused me much embarrassment. The putting off and on of shirts and trousers was perfectly sensational in its effect, and would no doubt supply capital material for the society gossips of Mahenge.

In taking my morning and evening promenade around the tent a surging crowd followed me, and necessary walks in the outskirts told very severely on my feelings. But as in the case of the lions at the Zoological Gardens, "the feeding" was the great attraction. A hush of expectancy would fall upon the crowd as the hour approached, and they watched with a feeling of awe the box being laid out and the camp stool set beside it, with the metal plate and cup, the bottle of salt, and the can of sugar, together with the knife and fork. I would then seat myself at this modest though not uncomfortable table. As the boy appeared with the stewed fowls and sweet potatoes, the excitement usually rose perceptibly and a crush for front places would ensue, threatening to upset my humble meal. The climax usually was reached when, with all the gravity I was capable of assuming, I took the knife and fork in my hands. The fowls, however, were leathery, and my unavailing attempts to cut and carve reduced the whole spectacle from the sublime to the ridiculous, and afforded such food for satire and laughter to the wags of the tribe that I blushed and scowled.

The day after receiving the chief's present, I prepared mine in return. On seeing it he expressed himself dissatisfied. I was not (he said) like an Arab trader bartering goods, but a white man travelling for his pleasure to see the country, therefore I should bestow on him something handsome. He had acted as my friend; for when the Arabs heard I was coming they dissuaded him from letting me pass, but he had not yielded to them! To this appeal I could not listen, and tried to turn the conversation to the subject of getting a guide, but without effect. I poured forth many grandiloquent arguments to show that I had given sufficient. I represented that if I went home and told my people that the chief of Mahenge detained strangers, refused to give them guides, and wanted large presents, they would not come again to his country; while if I could tell them that he is a great and generous chief and supplies his strangers with rice and goats in abundance, gives them guides, and sends them off in peace, then many white men might come bringing presents. This reasoning had some effect, and I was enabled to secure a compromise. I showed him my guns and revolvers, with which he was much astonished, opening his mouth, and putting his hand over it, as is the custom when anything surprises them.

In the evening I was treated to a war-dance. Only one man at a time engaged in it, dressed with all the barbarous splendours the savage mind can invent; huge feather head-dresses with tails, leopard and wild-cat skins, and a plentiful daubing of paint. Each performer's tactics seemed principally directed to drawing some supposed enemy from cover, and feigning retreat or escape from the spears or arrows of his opponents. Now he would slowly advance with fantastic prancing and curvettings, as if with all the pride and contempt of a Goliath before David. Next he would retreat carefully, covering his back with his large shield. Then amidst fierce volleys of encouraging shouts from the onlookers, he would make a sudden bound forward, jumping from one side to the other to escape missiles, make a motion as if stabbing, and finally shrink behind his shield and hurry back with doubled up body. Finally, to terrify and bid defiance to the enemy, he would put on the most outrageous faces, kick the dust behind him like an infuriate bull, shake his shield or strike it with his spear on his knee.

I shall now say a few words about Mahenge and its people. Strangely enough, though this interesting tribe is so near the coast, and though it is so notorious in slave-dealing, absolutely nothing was known to any European about it till we arrived at Zanzibar. The existence of some such tribe had indeed been heard of, but the people were supposed to be a branch of the tribe known as Maviti, discovered by Livingstone to the S. and S.W. of Nyassa. However, all was mere rumour and guesswork.

If the reader will refer to the map, he will observe the position and extent of Mahenge. It occupies a very acute angle, formed by the junction of the rivers Ruaha and Uranga, which constitute its boundaries to the north, east, and south, while the edge of the plateau marks its termination in the westward direction. Even of this small area the greater part is uninhabited, so that I certainly am not under-estimating the population when I put it at 4000. Yet this is the tribe which has been a standing terror to surrounding tribes, and which has depopulated tracts of country twice the size of its own. The wonder becomes greater when we find that in reality the Mahenge are as cowardly as the tribes they trample on.

Their language shows them to belong to the same race as the Wagindo, Wanindo, and the Wapangwa, who live south of the Rufiji. Originally they did not use any shields or spears, and doubtless were very much like the surrounding tribes. From all I can learn, the change in their character and position which has raised them into such a prominent place, seems to have occurred after a great raid of Maviti from the west side of Nyassa, who swept like a fierce tide across the plateau of Ubena and Uhehe into the valley of the Rufiji, and up to the very gates of Kilwa, which they nearly captured. On their return towards Nyassa a few of the tribe appear to have remained behind in Mahenge; the people of which they taught the mode of warfare practised by the Maviti, who, be it understood, are a section of Zulus who have settled themselves north of the Zambesi. So great had been the terror raised by these fierce warriors, that the very sight of a Maviti war-dress was sufficient to send the boldest to flight. The Mahenge having assumed the peculiar dress and arms of the Maviti soon found that they only required to show themselves to win an easy victory. And this they took such full advantage of that their doings raised a reign of terror, so that the helpless natives would not even muster to make any stand against them, and thus fell an easy prey. A bold face would soon have shown that it was a re-enactment of the old story of the donkey in the lion's skin. To this trivial circumstance of dress, the Mahenge are indebted for their present position; and this instance of the sudden rise of an important tribe in Africa is not without parallels, as we shall see further on.

Owing to the proximity of the country to the mountains, and the flat nature of the ground between the two boundary rivers, it is kept constantly damp throughout the year, and the soil being a rich alluvium it is one of the most fertile spots in Africa. All the cereals of the coast, such as rice (the favourite food), millet, and maize, are grown extensively. So also are such vegetables as sweet potatoes, yams, ground-nuts, melons, pumpkins, and cucumbers, and many other excellent articles of food. Tobacco is grown very abundantly. The sugar-cane, the castor-oil plant, and cotton, are also cultivated. Curiously enough in what appears to be an extremely favourable country for bananas none appear.

The Mahenge, as I formerly remarked, are a race evidently very superior to the Wakhutu. They are much lighter in colour, and have far finer features. The men have good figures, though many are somewhat effeminate in this respect. Hair on their face and head is also unusually abundant, luxuriant beards being by no means uncommon. The women are short and stout, few or none being good-looking. Many, however, have not the pendent breasts so characteristic of East African women.

The Mahenge dress in a somewhat varied style. The most common fashion among the men is a strip of cloth of the narrowest dimensions brought up over a cord round the waist before and behind. The women have not the cloth continuous, but in two pieces, each about one foot square. Bark cloth and pieces of skin are also frequently used. Those rich enough to provide a good-sized cloth wear it thrown over their shoulders, instead of utilizing it as a loin cloth, which would be the more decorous style. A dress is frequently made by cutting monkey-skins into narrow strips, and then twisting them till they appear like so many cat's tails. These are attached all round to a leathern belt, and thus form an airy kilt.

On the war-path cloth is entirely laid aside as too effeminate, and they issue forth with immense feather head-dresses. Cat and leopard skins are donned on the back, breast, or sides. Circlets of hair are tied round the forehead. These, hanging down over the face, give them a very savage appearance. They wear few ornaments,—triangular pieces of shell strung together to form a necklace, being used by the men, while the women have armlets of thick brass wire. The hair is not dressed, but plentifully greased with castor-oil.

The Mahenge present also a marked contrast to the Wakhutu, in the want of respect for their dead. The latter bury them decently, keep the brushwood clear of the graves, and usually erect small pent-houses over them, where they place offerings of food. The Mahenge have no such customs. They tie their dead up in the crouching position which they love so dearly during life. They are then put in a hole in the jungle, and no mark left. Hence no sign of graves is ever seen.

The women have a curious mode of expressing grief at the death of a relative. They actually dress themselves in ropes and cords, and sit at the doorway for several days.

The people do all their cooking out of doors, a very unusual custom in East Central Africa. A peculiar red clay with mica in it, and called likite, is eaten to a considerable extent by the women. Such a craving have they for this, when once the habit is formed, that notwithstanding the dreadful effects of the practice they are quite unable to discontinue it.

The houses of the Mahenge deserve a word of notice from the curious variety of styles which they exhibit. Some have walls of wicker-work, looking like huge circular or quadrangular baskets with roofs. Some are of the common square form, but instead of the walls being built of poles and mud they consist of only a few poles, the rest of the wall being made up of a double thickness of millet stalks, the outer layer placed perpendicularly, and the inner horizontally. In such cases the house is frequently put up in sections, each side being first made, then stuck up. These houses are easily removed from place to place, according to the pleasure of the owner. The most peculiar and unusual style, however, is that in which the house is built on a platform three feet from the ground. The form is circular, and the roof is made so large that it projects down all round below the level of the platform, so that, when standing outside, nothing but a huge cone of thatch is seen, the walls being hidden from view. In such a damp country as Khutu this form of house presents many advantages, raising the floor out of the wet during the rains. Many of the huts are of the usual Ukhutu form.

The Mahenge set apart separate huts for their sheep, goats, fowls, and pigeons, while their pariah dogs have to shift as best they may. Separate huts, raised high on poles, are used as granaries.

After our exhibition of four days, the chief graciously gave us guides, who required to be paid at exorbitant rates, and allowed us to depart. Making a détour, to avoid the lagoons and backwaters of the Uranga, we marched N.W. through jungle, with scattered hamlets and patches of cultivation. There was nothing to relieve the monotony of our surroundings but an occasional fan-leaved palm, which in the breeze produces a sound like falling rain. Crossing the inevitable swamp, we camped at Gambula. Here we were made the victims of some "artful dodgers," who brought us four baskets of rice and two fowls, representing them to be sent as a present by the chief. I, of course, made a suitable return present of double value, to find afterwards that there was no chief, and that we had been imposed upon.

A fit of ague kept me cool for two hours in the afternoon, and I got no sleep during the night. Setting off as usual with the rising sun next day, we rounded the deeper parts of the marsh, and, crossing a considerable stream called the Dete, by means of the fallen trunk of a Borassus palm, we reached a small village called Pédete, where we rested. We heard that two slave caravans had bolted into the jungle on hearing of our approach, expecting that we would attack them, and set the slaves free.

On resuming our day's march, we were soon stopped by a great outcry among the villagers. A porter had stolen a cooking-pot, which he had got the use of. The caravan was at once halted. The thief and his confederates were found, and both severely fined and flogged.

Here we observed the method by which the Mahenge obtain salt. They burn certain grasses, water-plants, and trees, the ashes of which are filtered through a cone of basket-work, lined with leaves and filled with grass and mud. The salinated water is then boiled and evaporated, leaving a very impure salt behind.

At Paliogoalina, where we camped on the third day from Mkomokero, we were very much amused by the old chief of the place. Bringing a really princely present, he sat down in the usual doubled-up position, and assumed a most stoical expression. For ten minutes he sat without moving a muscle or uttering a word, leaving his son to speak. At last he rose, assumed an oratorical attitude and tone, and waved his stick. After fairly impressing us by his appearance he made a capital speech, telling us how he had once been chief of all Mahenge; but when he got old he retired to his present place, leaving the government in the possession of Komokero. He was much interested in some photographs, and the hoary-headed old sinner wanted to know how he might get a few of the ladies as his wives.

At Pakechewa, where we camped next, I determined not to wait for the chief's present, but to give mine first, which would be an immense saving to us. I accordingly sent off to him a small present. It was returned promptly with the remark that it was not the custom of the Mahenge to receive presents from their strangers, but that he would come in the evening and talk over it. He did arrive in the evening, and proved to be a fat, intelligent personage, from whom I received a good deal of information; but as I never got the same story from any two persons, I had learned by this time to put no faith in anything I was told.

After leaving, he sent to say that I must not go next morning, as his people wanted to see me and learn the difference between a white man and an Arab. He hinted that if I tried to go on, he would consider I was forcing my way through his country, which would be equal to a declaration of war. I however determined not to be stopped in this absurd way. After a very bad night with rheumatic pains, I rose, amused at the idea of a petty African chief stopping me. Going out, I was surprised to see not the slightest sign of preparation for starting. On inquiry, it appeared that the guide dared not and would not move against the chief's order. I tried threats and promises, but without effect; and, as the men would not support me, I had, as in previous instances, to resign myself to the stoppage. I had some bird-shooting during the day.

In the evening I had an unpleasant insight into the temper of the men. Seeing a porter offering beads of a kind suspiciously like my own to a woman to pound rice for him, I asked Chuma where the man got them. The latter, who heard what I said, immediately went off among the men, telling them I was accusing them of stealing beads. A dreadful row was at once raised, the drums were beat frantically, and the horns blown to call all the men together. From all sides they came rushing, bringing their guns, &c. These they laid down at my feet with the air of injured innocence. They had never been accused of stealing before! "Here are our parcels," cried they; "look and see if we have anything belonging to you Now give us our tickets of discharge, that we may go back to the coast, for we cannot go with you to be looked upon as thieves!" Every one was in the utmost excitement. Personal articles were packed, and preparation for a general return made, as if an unpardonable aspersion had been cast upon their unsullied honour, which as immaculate men they were bound to resent.

I explained what had really occurred, and represented that I had accused no one of stealing. But my explanation had no effect. Then I became apologetic, and appealed to them to remember I was but a boy, wholly inexperienced, and therefore liable to make mistakes. They should be fathers to me, and tell me quietly and gently when I went wrong, so that I might be put right; and not rush off wildly beating drums, throwing down their guns, and declaring they would return. I was but their pupil, and that was a bad way to teach me how to do right and learn to travel with them!

This harangue took immediate effect, and they became so enthusiastic in my favour, that they at once commenced a dance of universal good-will. I was more careful of touching their tender feelings of honour in future.

We were here startled by strange news. A white man's caravan, with a woman and four elephants, were said to be close behind us, and trying hard to overtake us. This was how the news of the start of Captain Carter's elephant expedition reached us away far to the south of his route. But as I was ignorant of the road he was going to take, I naturally supposed that what I heard of their following in our footsteps was true. My national pride revolted at the idea of a Belgian expedition overtaking and perhaps passing us. The men were at once convened, and the business stated to them. Such a thing they declared they would not endure, though they marched till they fell on the road. They were ready to start in the middle of the night. So great indeed was the enthusiasm, that they turned the place into a very pandemonium, forming a procession round the village, firing guns, and shooting, thoroughly alarming the entire district.

Next day after satisfying the chief with a present, we resumed our march. An hour's walk brought us to a fine stream or river, called the Luipa, which had to be crossed by canoes. We camped on the opposite side after getting everything safely over. Here our Mahenge guides made a demand for more cloth. On being refused, they deserted. However, we had no difficulty in getting other two.

The next day's march would bring us to the base of the mountains which rose grand and majestic before us. As we started in the morning, we anticipated the clear streams and invigorating air of the higher ground with much pleasure. After two hours, through most harassing sedges, we entered a tract of forest which proved to be a piece of the most pleasant walking we had experienced for many days. The trees were large, and covered with huge tree-like parasites and creepers of every description. There was little undergrowth, and the paths were open. We soon, however, re-entered the jungle, descending into the basin of the Loandwa, a considerable stream flowing to the Luipa. Crossing it, we passed through patches of woodland, jungle, and cultivated fields, reaching after a long march the village of Mmatanga, charmingly situated on a wooded knoll. To the north-west of the village extends a pleasant ridge, covered with dark green trees. Behind it another and a higher ridge rises, while beyond, the Uchungwe Mountains tower aloft, half veiled in drifting clouds.

CHAPTER VI.

THE PLATEAU OF INNER AFRICA.

THE reader can hardly be expected to understand the sense of exultation and delight which I felt on reaching the end of our tramp through the lowlands of East Africa. It was as though a great and decisive battle had been fought and won. Africa has been compared to a nut, only hard to deal with from the outside. Once through the shell, and the prize is gained. We had got through the shell, which in East Africa means the low-lying country between the coast and the edge of the Plateau. This is in many respects the most trying part of an explorer's journey; for here the European is first brought face to face with the hardships of travel; here he has to do battle with disease, bad fare, hard work, and danger, and to be ever on the alert against desertion and stealing. Nature soon weeds out the incapable, and determines who is "fittest to survive."

Few caravans have crossed this tract of country without manifold troubles, as the history of all

East African travellers has shown. We read of nothing but marshes and swamps, fevers and dysenteries, difficulties with the natives and with their porters, desertions and cases of stealing, until utter ruin seems to face the explorers. We too, had had our share of most of these things. Our leader, who seemed to be so admirably suited for the work, both mentally and physically, had fallen a victim; and a gloomy fate seemed to await the Expedition, as I myself had nearly succumbed to fever, and the caravan had been almost scattered by Mahenge. Troubles and adversities had beset our path. But we passed triumphantly through all with one great fact to boast of, and give us renewed strength and encouragement—we had not lost during our march of over 350 miles a single man by desertion or death, and not a yard of cloth had been stolen.

No caravan, whether Arab or European, had ever performed the same feat; in this our experience was unique. Our progress had been slow, but it had been sure. We made no brilliant march, but we did something better. We passed in peace through every tribe, leaving nothing behind but good-will and friendship. We taught the natives that our mission was peace, and that the word of the white man could be trusted.

To the crossing of the high mountain ranges which bound the interior Plateau we now braced ourselves with energy. We learned that three days would be required to reach the Plateau, on two of which no food could be got. Two days' food was accordingly without delay provided, to be carried for use, and on the day following our arrival we commenced the ascent with buoyant spirits. I already sniffed the cool fresh mountain breeze, and felt as if it were making my blood course merrily through my veins. But for the fact that it would have been somewhat undignified in the "Bwana mkubwa" (great master) to whoop or dance, I would certainly have done so, weak though I was from recent fevers.

Crossing a low-wooded ridge my hilarity was somewhat damped in a miry swamp, which I commenced to cross on a slippery fallen tree with much coolness and indifference. Before I had got half-way across, I suddenly disappeared in a pool of liquid mud, and required the assistance of half a dozen men to drag me out, half drowned. After being scraped down I continued my march in the most sloppy and bedraggled condition. The odour ascending from my clothes kept me in a more subdued frame of mind for the rest of the day.

A steep hill now presented itself before us, with the footpath leading quite straight up; no zig-zagging about, taking advantage of this place and the other to make the ascent easy, but direct to the point at once. Commend me, however, to the Zanzibar porters when a piece of hard work is to be done. There are no demands for extra wages, and no discontented grumbling. They get their drums to the front; the horns are blown, and the men sing and shout as if commencing a dance. That was the way we set to work. The drum, vigorously thumped, sent forth its volume of sound, accumulating in force till every hill and glen seemed to roll and thunder. The notes of the horn, loud and harsh when near, echoed and re-echoed on all sides, coming back to our ears from the distant hills with a plaintive cadence pleasant to listen to. Higher and higher the drums got above us, inviting us onward. Now the Kiringosis would start a resonant recitative, to be answered by the men far below. Anon Chuma would shout some cheery remark to the men in front, to be answered with laughter. Those who lagged behind or sat down to rest were greeted with volleys of ridicule worthy of a London cabby or omnibus conductor. Every one tried as it were to beat his neighbour. The power of lung and muscle displayed by the Zanzibar porter is certainly remarkable. With a load of from sixty to seventy pounds on the head or shoulder, and a gun in the one hand,—the other being occupied steadying the bale,—he will patiently toil up a precipitous mountain by the hour together without stopping to rest, and probably shouting or singing all the time.

As we thus continued to ascend our prospect widened and extended, till we obtained a bird's-eye view of the whole country we had traversed.

Away to the N.E. we could just descry the Duthumi mountain on the horizon, with Mount Johnston a little further east. Between lay the wearisome plain, with its yellow or dark green colours, indicating the jungle and the forest. Here and there a silver thread could be traced winding about, indicating the course of the stream. Around us was a billowy sea of hills, rising in higher and higher waves, till the culminating heights of the Uchungwe Mountains were reached.

A stiff pull brought us to the top of the lower flanks, and after surmounting a number of minor ridges, we camped at a small hamlet. The next day was a very short march, owing to impending rain and the length of the march to the next station.

I now began to be aware of the evil effects which the long residence in the malarious lowlands had had upon my system. The poison began to show itself, and as it were to ooze out under the effects of a purer atmosphere. It seemed to take the form of rheumatic fever, and acted with such dire effect on my lungs and heart that I thought it would kill me. The hard work of an entire day seemed concentrated in each step I took. Every few feet I had to stop, gasping for breath and blowing like a broken-winded horse, while my heart palpitated in the most alarming manner. My mind became dazed and stupid, while my poor limbs seemed made of jelly, so utterly incapable of supporting me did they feel.

In this dreadful condition I continued my journey,

wearily dragging myself behind the caravan, and struggling desperately to keep up with it, but finding myself each day gradually falling back till I would be left alone with my servant in the lonely solitude of the mountains, to wonder where all this would end.

Yet ever and anon, as I sat down to recover my breath and stay the palpitation of my heart, a feeling of strange fascination would steal over me while I surveyed the grand panorama. All around rose the massive mountains in peaks, and domes, and serrated ridges; here covered with a grass-green mantle, there varied by a darker shade of noble forest. The distant sound of a falling cascade, or the softer and more plaintive sough of the stream from the valley beneath, laughing and babbling over its rocky bed, was brought as a long-drawn sigh to the ear by the mountain breeze. Now and then the caravan could be descried in the distance, winding like a great snake by rocky dell or gloomy valley, ascending this ridge or rounding that peak, now appearing on an open glade, anon disappearing in a clump of forest, while a picturesque group of absolutely naked savages crowned the neighbouring heights, faintly discerned through the driving mist, and watched with surprise and apprehension our movements. Then how keen was our delight to find by the rocky banks of the streams such old friends as the bracken bringing to us as it were a breath of our native air! Under the shadow of that prince of ferns, the tree fern, grew the well-known royal fern, and in the shade of tropical trees and shrubs in rocky crevices nestled a species of maiden-hair, rivalling anything African in its tender grace and beauty. These were charms which no amount of fever could blind us to. They acted as a solatium to all our troubles, and fascinated us with a boundless sense of liberty.

On the third day, after a very hard and wearisome march, we camped beside a solitary house on a bare, exposed ridge, with magnificent scenery on every side. There was no shelter for the men, and they were consequently compelled to lie down with no other cover than their thin cotton sheets. The cold season had now set in, and the altitude we had reached was nearly 4000 feet, so that the temperature was very low. During the night a frightful tempest of wind and rain came on, and in our exposed situation we got its full fury. Wet and shivering we had almost to cling to the ground to prevent ourselves from being swept off, and thus we lay throughout that terrible night. When day broke, the light revealed a melancholy sight of men lying huddled together for warmth and protection, and seemingly half dead. Afraid that some might be wholly so, I roused every one up, and much to my relief found all alive, though they were so stiffened by the cold as to be almost incapable of motion. The storm still continued, but to keep the men from succumbing to its effect I compelled them to move about and get some warmth and life into them.

This, however, was not our most serious trouble. We now learned the fact that for some purpose or other we had been directed the wrong way, and that instead of taking three days to cross the mountains, it might be double or treble that time. Meanwhile, where were we to get the necessary food? No villages existed, and only a solitary hut and small cultivated plot occurred at wide and unknown distances. Here was a pretty problem to face! However, we determined to proceed, and if it came to the worst, then the word would be "*sauve qui peut!*" Leave everything behind, and each man act for himself! On that day there was no fuel to make fires, and though it still continued to rain all forenoon, the men accepted their fate with much resignation, and ate the little uncooked food that was doled out to them.

At midday the violence of the tempest ceased, and it became necessary to resume our route. Descending a rocky and almost perpendicular pathway, with many treacherous slippery places, we reached the bottom of a deep gorge, nothing worse having befallen us than a few awkward falls. Through this gorge we wearily found our way, jumping from rock to rock, tripping over unseen creepers, and crossing and re-crossing the swollen streams. The high mountains on either side rose in perfect precipices, but wherever a footing could be obtained superb tropical trees arose, reminding us of the many strange forms we met in the grand forest of Usambara. Tree and other ferns graced the banks of the streams, and at every step some new novelty greeted our eyes. Late in the evening I crept into camp far behind my heavy-laden men. There was but one house lying at the bottom of a steep mountain. Our prospects looked gloomy indeed, as we distributed the last of the food; less than half-rations for each man. During the night the rain once more descended in torrents, though fortunately there was no gale.

In the morning I felt as if I was incapable of moving a step; but it was a question of going on or starving. So up the mountain side I struggled, trying to look as cheerful and hopeful as possible before my followers. The men were hungry, but no one would have inferred there was anything wrong on listening to their cheery shouts as they went manfully to their work. The top of the mountain we found covered with a fine forest; the trees rising straight and unbranched to a considerable height, before spreading out in a fine umbrageous canopy. But their chief beauty lay in the astonishing abundance of a grey-man's-beard lichen, which hung in silvery hairy masses from every twig and branch, covering the trunk till it might have been imagined the trees were there only as exhibiting posts for this curious plant. Its abundance is doubtless due to the constant dampness of the atmosphere, the rain-clouds from the Indian Ocean leaving here a considerable portion of their burden in the passage to the interior.

We had now reached a height which made the climate temperate, and no better evidence was required than the presence of the numerous well-known genera of plants. Indeed, there was little in the scene to indicate our position in the tropics, every sight possessing such a familiar aspect that, if it had been transplanted into England, there would have been little abnormal to be noticed, at least by the casual observer.

After reaching the top, we commenced our descent of the other side. At the bottom we found many royal ferns beside the stream. After a rest I was pulled half way up the opposite side, where we camped. A small goat was divided among 150 men, as the only food to be got, and strange to say there was next to no grumbling. The Zanzibar porters, who have so frequently been vilified till they have appeared the very incarnation of all that is bad in man, acted like heroes. They saw at once that I had not knowingly led them into this pickle, and that we were the victims of the people of Mmatanga. They accepted their position as one of the misadventures a caravan through an unexplored country was liable to fall into, and they made the best they could of their circumstances. Though there are few people who have a keener delight in gorging themselves with food, yet they have a power of enduring abstinence which is truly astonishing.

When on the march, they never under any circumstances take more than one meal in the day. They start with empty stomachs, march all day with them empty, and only begin to fill them towards evening. Curiously enough, whenever they remain a few days at any place doing absolutely nothing then they think their fate is very hard if they do not get three meals during the twenty-four hours. I observed it to be invariably the case that, when working, one meal sufficed, though they might have had more; while, when they were idle, they would have three.

Our sixth mountain march was not so harassing as the preceding. Our route led diagonally up the mountain sides, then along two ridges. Finally we camped on the top of a mountain at a height of 5000 feet. The men were at once set adrift to search for food. For this purpose they explored every mountain and glen for miles around, returning with scant reward for their labour. A little maerè grain (*Panicus spicatum*), and another grain, called by the natives ulizè, together with a few sweet potatoes, was all they obtained, and it did not one-third suffice for a meal. The men were drawing very long faces, and beginning very decidedly to feel the bottom of their stomachs. Immensely to our delight a bullock was brought into camp for sale, which after much haggling we bought at a high sum. This raised our spirits greatly, and Ngombe, our butcher and caravan-bully, was proceeding with great alacrity to slay; when to our horror and dismay, with a wild bound it got free and went at a tearing pace down the hill. A howl of dismay rose from the astonished men, and they seemed to have lost the use of their limbs, as they saw their anticipated and much-needed supper escaping. Seizing a rope's end I brought them to their senses, and cleared the camp in a twinkling. Then a wild rush of men took place down the mountain side. Slipping, rolling, and falling, on they went as if for dear life. The bullock kept ahead, goaded on by its pursuers. It reached the bottom, crossed the hollow, and became lost to view in the forest beyond. Darkness came on, and I strained my eyes in vain to follow the pursuit. Faintly now and then came rolling from a distance the shouts of the men as they directed each other. For three hours the pursuit lasted, and I gave up hope, but much to my relief triumphant shouts were heard far beneath in the valley, and soon they arrived with the bullock securely roped, giving evident signs of a hard chase.

Next day, on taking the men into council, they all voted for staying where we were that day, to give them time to make a long foraging expedition. Though I thought the scheme very hazardous in our critical position, I let them have their way, and fortunately after dark they returned with half-rations for two days. To this we had added another bullock. In this way we contrived to keep life together four days more, as we ascended and descended hills covered with a remarkable assemblage of flowering shrubs, which filled the air with an overpowering honey-like fragrance.

On the 17th of August we marched over a pass 7000 feet in altitude, and found we had at last reached the Plateau of Inner Africa, leaving the mountains behind us. Next morning our brave little band resumed its march with much alacrity. Pushing our way down a low valley we ascended a hill from which the grass had just been burnt. Before us appeared a succession of low rounded hills covered with short yellow grass, dotted here and there with deep green clumps of what appeared to be trees, but which afterwards proved to be only shrubs. Ascending another hill the view became extensive. In the far distance rose the apparently even hills of Ubena, passing northward into the more broken outlines of Uhehe. Between the ground rose and fell in wavy undulations, smooth and rounded, with not an irregularity to relieve the scene. A more bleak and barren prospect as far as the eye could reach could hardly be conceived. And yet this was Central Africa, and in the tropics! As we stood there on that commanding hill a great disappointment stole over me. I indeed felt that we were conquerors, and that the country was at our feet. But was this desert worth having? Did it repay the great sacrifices we had made to reach it? These were the thoughts which first passed through my mind on scanning the wide uninviting landscape.

At midday we reached a Wahehe village (for Uhehe is the name of the country we had now entered), where for the first time we saw the *tembe* form of a house, with its flat walls.

When we arrived, we all expected to feed on abundance; but much to our dismay, little grain was to be got, and the natives absolutely refused to sell us any bullocks, as they declared they belonged to the chief, and could not be sold without his permission. Though we were ravenous with hunger, it would have been disastrous to our future movements to enter a new country with high-handed measures. So we left the Wahehe to please themselves. They evidently were extremely suspicious of us, and of our intentions. On the following day we despatched messengers to the next village to announce our approach.

When we came within sight of Mwhanna, we were surprised to see numerous fine herds of cattle rushing wildly across the country, driven by naked herdsmen. Women were seen hurrying along in twos and threes with their children, and bundles of what appeared to be their property. Armed bands were congregated about the villages. Thinking that possibly some neighbouring tribe had made war on them, we hurried our movements. But a large stream called by the Wahehe the Ruaha, but which was only one of the Ruaha's tributaries, lay between, and had to be crossed. As it was very deep, and the bridge which spanned it was of such a dangerous and suspicious character, that I would not trust my bales on it, the "Agnes" had to be got out. With marvellous celerity, our trusty boat conveyed everything safely over.

Before we had finished, we were relieved to see our messengers return, about whose safety we were somewhat concerned. We then learned that we ourselves were the cause of all the commotion we had seen. As we had come from the mountains unannounced, they had concluded we were an Arab caravan, who had joined some rival chief with whom they were at war, and had come to attack Mwhanna. Hence the cattle were driven off, and the women sent to the neighbouring mountains, leaving the warriors free and unhampered to fight or flee as circumstances required. On my men appearing without warning, they were at once seized, and for the moment were in considerable peril of their lives. But when it became known that it was a white man's caravan which was coming in peace and friendship, the panic was somewhat allayed, though the warriors were now rather at a loss what to do, as the headman of the village, who is the son of the chief of Uhehe, had fled in dismay to his father. However, we might now come in safety to the village and await further instructions there.

For the first time in three months I had some fresh milk to drink, and, getting a present of a bullock, we feasted and made merry, though hardly to repletion, as no grain was to be got.

Mwhanna formed one continuous line of flat-roofed houses, divided longitudinally into two parts, the front and back. The front part is occupied by the cattle, and remains undivided throughout the greater part of its length,—all the cattle being housed in common during the night, when they are carefully shut up without food, to prevent them being carried off. The back part of the house forms the dwelling-place of the Wahehe, and there they store the milk. It is partitioned across to suit families. The entrance is through the cattle's portion, by which they also get the little light they require, and as there is no other means of ventilation, the natives are kept warm by the crowding and hot breathing of the cattle,—not a very wholesome way, certainly; but the natives are by no means fastidious. They never light fires or cook in the back rooms.

The day after our arrival all the women returned, and we were enabled to scrape together a little food at high prices. To prevent any more surprises, Stamboul was despatched with the under-chief to the chief of Uhehe, to announce our approach according to custom. He carried a handsome gold cloth as a present.

Next day we followed our ambassadors, tramping westwards along a shallow valley, through which a stream winds to the Ruaha. A considerable number of house-rows and herds of cattle testified to the grazing capacities of the ground. The marked absence of cultivation was a subject of astonishment to us,—small garden-like plots from which a poor crop of melons and ulizè is raised, being the only sign of it round the village. Evidently, from their scarcity, these articles are regarded as luxuries not always to be indulged in. Of vegetable food the melons seem to be the most important, and large heaps of them could everywhere be seen stored on the top of the flat roofs. Firewood is hardly obtainable, and the natives have to utilize the dung of their cattle instead, first drying it into bricks, and then storing it up—a very necessary practice, as they are not infrequently besieged in their villages by war-raiders. Cattle-dung is also used to mix with the clay with which they build their houses. We saw some quagga during the march, but the country was too open to permit of our approach to them.

The appearance of the country, as first observed from the hills, was not belied by a nearer acquaintance. Everywhere it presented the same uninviting and monotonous appearance, rising into smooth rounded ridges, to fall away into shallow rounded valleys, with the same green patches on a ground of yellow—a row of houses on the face of that knoll, a herd of cattle in the hollow, here and there a naked Mhehe, and in the far distance a herd of wild asses or quagga. Overhead circled a tawny, ragged vulture, and on a rock sat a number of crows with white necks. Not a sound breaks the stillness. In Africa animals seldom make a noise. The cattle do not low or the dogs bark—doubtless an inherited instinct in a country where lions, leopards, and hyenas abound. Such were the disappointing characteristics of the scene we now traversed.

On the third day from Mwhanna we reached the village of the second most important chief of Uhehe. Here we were compelled to stay two days trying to get guides. We gave the chief what we thought was an excellent present, which was very graciously accepted, and guides were promised. In the afternoon news came intimating that no one could be got to show us the way. The chief was very sorry, but he could not force his men. This was disappointing enough, but there was worse behind. The harvest had been so bad that he could not give us a suitable present, but he sent a bullock instead. The latter did come, but we had just time to see it before the wild brute made an irresistible rush, and took to the plains, and we never set eyes on it again. In the evening I sent to the chief for some milk. Much to my astonishment I only got excuses. "None of his cattle were giving milk, and, though he had tried hard, his people would not part with any of theirs."

A new light broke in upon me. The chief was evidently not satisfied with the present I had given him, and took this gentlemanly way of breaking the fact to me. I therefore sent him an additional present to soften his heart, accompanied with a request for milk and guides. Much to his surprise he found that, after all, he could get some milk for me, but he regretted to say that his people would not move as guides. Clearly he was still not satisfied. On the following day I had an interview with the great man; and on my reasoning against his greed and want of hospitality, he laid down his argument as follows:—"The Arab traders come up from the coast with small caravans of six, seven, or eight bales of cloth to buy ivory. They make me (the chief) a present of nearly half of these, and yet when you come with a large caravan you only present me with so many cloths, even when you want to buy nothing from us." It was useless to argue with him, and as it was impossible to move without guides, there was nothing for it but to make an addition to my present. At last he was satisfied, and was delighted to discover that even his own cows now gave milk, and that two of his men would be most happy to accompany me as guides.

We found a small trading caravan at Misimiki, which had come by way of the old Usagara route. We took the opportunity of sending letters home with them as they were on their way back.

The Wahehe here were seemingly very well off, and it was extremely interesting to see them moving about in groups, carrying long wands, with their tall, well-made figures classically clothed in a huge extent of white or blue cotton, which fluttered in the breeze. This was thrown on anyhow. Decency in dress was a thing unknown to them, as they thought nothing of stripping themselves stark naked in a crowd. When on the march or doing any work they never wear a single rag. The men are passionately fond of blue beads, of which they wear pounds' weight strung round their neck and breast. The women are equally fond of these ornaments, but have to be content with a less quantity. The dress of thé Wahehe damsels is by no means conventional or straight-laced, considerable latitude being allowed both in the extent and character of it. Those who can afford it dress almost entirely with beads, wearing innumerable strings round their loins. Nothing else is worn in front, while, behind, it would be considered indecent to appear without a piece of hide ornamented with beads, and cut away so as to hang down like a tail. Doubtless this latter idea of the requirements of modesty has been derived from the observation of the animal kingdom, the four-footed members of which are so markedly provided with a caudal appendage. This, though a somewhat absurd idea, is probably more feasible than the theory that the custom has arisen from the attempt to supply an artificial tail on the loss of an original natural one at some early period in the history of the human race.

The scantiness of the Wahehe dress, however, is more remarkable when the extremely trying nature of the weather is considered. Their covering of beads, though somewhat heavy, is by no means wind proof, being on the contrary exceedingly airy.

The love of adornment, and the wish to set off a good figure to the best advantage are the only reasons that can be assigned, and are only another illustration of a universal law.

Guides having been at last provided for us, we resumed our march with pleasure. At Litungirwa, where we camped for a night, we were much pleased to see some rude attempts at adorning their houses. The tembe is here built in the form of a square, into which the cattle can wander from their sheds. The walls are neatly plastered with mud, on which drawings of different animals are painted in white kaolin. At one place a snake and a crocodile were carved in relief. The doorways are in the form of a heart with the apex down.

On the second day, while tramping along, we were surprised by a messenger from Mamlè, the head chief of Uhehe, informing us that he was very much displeased at our want of courtesy in breaking their customs, by not sending a messenger to inform him we were coming. Having thus wounded his dignity, we were refused an interview with him till it was his pleasure to send for us. He was going to another part of the country, and meanwhile we were to proceed to the chief's head town, there to await his coming. Here was somewhat startling news! We were in the heart of the country, and indefinite detention seemed to await us. Powerless to act, we could only await the course of events, and thank God we were at least going forward. I at once explained how we had arrived, and had expected that everything was duly notified to him. I also declared our motives for traversing his country.

A short time afterwards, seeing the messenger speaking to Chuma, and thinking he had gone and come back, I began to shout angrily for them to come at once to me. On hearing me the messenger at once took to his heels with fright, and was seen no more, and I learnt that he had never been away; and now he was likely to convey some extraordinary report to Mamlè about my ferocity and dreadful manner. Rather anxious about the turn of events, we marched on to camp.

The old saying, however, that misfortune never comes singly, was about to be illustrated in a very alarming fashion.

From the day in which I assumed the leadership it had been my constant endeavour to infuse a kind of *esprit de corps* among my men. I took them into my confidence, asked their opinions, and got them to narrate their wanderings. I pointed out how differently and how much more successfully we had got on, compared with all other caravans. I hinted how they would be able to boast that no one had deserted, no one had stolen, and that they had been the first who had ever crossed these countries. To such words they listened with enthusiasm, and began to feel that they had a reputation and a character to keep up. They had an unbounded admiration and regard for Dr. Kirk, and I took advantage of the share our accomplished representative at Zanzibar had in organizing the expedition. I tried successfully to make them believe that Dr. Kirk had handed me over to their charge, to be carefully conveyed over Central Africa, to be shown the sights, and brought safely back to the Baluzi or ambassador again. This notion took admirably with them and raised their self-esteem immensely till they really believed they were conducting me, in place of me leading them. By judiciously fostering this notion I got a command over them which was not otherwise attainable. And now it would have been felt as a disgrace to the whole caravan if a man had either deserted or stolen.

In pursuance of this scheme I tried to abolish flogging for misdemeanours, and to substitute fines. I told them about the degradation attending such a punishment according to our notions. Against this innovation, however, they grumbled exceedingly. They were accustomed to being flogged, they said, but fining they knew nothing of. A flogging lasted only for the minute, but were they going to travel so far and come back to find their money all fined away! Such a thing they would not hear of; and as I persisted, much mutinous feeling was the consequence.

There were other reasons, however, to make them rebellious. For a whole month they had never got sufficient rations, and what they got was very bad and indigestible. To add to these grievances a very large proportion of the men were very ill with hæmorrhage, influenzas, cramps, and rheumatism, the result of the climate they were enduring. From morn till night a cold east wind swept with freezing fury across the bleak and shelterless moorland, making the poor fellows shake and shiver as if in a snow storm. At midday, and under a cloudless sky, we have been actually compelled to halt and light large fires to warm the benumbed porters. I have marched frequently all day long, dressed in tweed clothes and wrapped in an ulster; and I was from a cold climate! What then must the men have felt coming from the hot and equable climate of Zanzibar, and dressed only in a thin cotton shirt? There was little wonder that a spirit of insubordination should arise and find vent in some unpleasant manner.

On the 28th of August, the day after our alarming message from Mamlè, we had stopped beside a stream to light fires to warm ourselves. After an hour's halt, I thought it time to move on to the village at which we proposed camping. The men, however, were warm and comfortable, and liked not the idea of leaving the fires. On my giving the order to march no one moved, and every one seemed to say, "We are better where we are!" In a moment of irritation I seized my empty vulcanite water-bottle, and, swinging it round my head, I spared no one who came within my reach. With surprising rapidity our camp was cleared of every man, with the exception of two more dogged than the rest.

Knowing they would follow, I intimated to them clearly that I should fine them so many rupees for their insubordination.

As I expected, they arrived in the evening, and forthwith commenced to instigate the rest of the porters to resist their fining. Both were influential in the caravan, and they got only too many ready ears to listen to them. At sunset the porters came in a body, and threatened to desert to a man if I did not remit the fine. I was nettled, and would not listen to them; moreover, I never believed they would carry their threat into effect. They all retired, but in a few minutes returned, defiled past me, laying down their guns, and the Kiringosis even their red cloaks at my feet. Everything, with the most scrupulous honour, was brought and deposited. This looked serious; but I was determined to remain firm, and see how far this matter would go. During the night the camp was in a state of commotion; everything was got ready for a march to Unyanyembe, and the routes to it were discussed with an acuteness and an amount of knowledge for which I had never given them credit. It may be well understood I slept little that night. But I was ill and obstinate, and resolved to fight to the bitter end.

At daybreak the drum beat as usual, and the barghumi or horn blew the customary blast. I went outside and sat down. With indescribable feelings I watched the men. Their reckless talk and laughter indicated the excitement they were in, as they ostentatiously tied up their bundles and prepared to start. At last everybody was ready. With a wild defiant cry they defiled out of camp. Some ran, some sang, and in a few moments all that remained were the six headmen, the cooks, two boys, and the one porter who had his wife with him. With these I was left in the heart of Uhehe. An hour before, the prize of Lake Nyassa seemed surely in my hands; and now utter ruin, and even death itself stared me in the face. But what about the position of the porters who had thus sacrificed themselves to support their two comrades? They dared not return to the coast. The only other place they could go to was Unyanyembe, where at least they would be safe for a time. But how were they to get there? Three hundred miles of country lay between, inhabited by the fiercest and most warlike tribes in East Central Africa. They had no cloth to buy food; for not an article belonging to me had they taken. They had no weapons to take food by force, or defend themselves if attacked. Clearly death or slavery would be the consequence if they persisted in their attempt; and all this had been risked for a mere trifle!

For me in this crisis there was clearly but one alternative—submission or disaster. Wild schemes of setting the Wahehe warriors after them rushed through my head, but were at once abandoned. At last, in an agony of despair, I turned to Chuma, and said that I must give in to their demand. Chuma instantly went after them with the news, but they, perceiving that the game was now in their own hands, affected never to mind him, and kept going on. Seeing this, and getting thoroughly frightened, I also ran after them, and energetically declared that I would never fine them again. If they would only return to their post, I would henceforward flog them to their hearts' content. This promise put everything right at once. Triumphantly they re-entered the camp, shouldered their loads as if nothing had happened, and marched off with drums beating and horns braying.

It has, however, always since been my opinion that the whole was a piece of bravado on their part, and that after all they never really contemplated deserting, knowing full well that I would submit to worse things rather than let them go. But I never forgot it, and a leather belt which I wore was wielded without compunction till it became a standing terror to evildoers. A significant motion of my hand towards it usually indicated that my patience was exhausted, and that, words being ineffectual, action must be tried.

The day following this exciting episode we reached Mkubwasanya, the town of Mamlè. This proved to be a miserable collection or row of houses of the most rickety construction, and situated in the most dreary waste we had yet seen. There was actually no cultivation and no firewood for miles, and the place being high and exposed the cold was intense at night. The variations in the temperature were most trying. At four p.m. I have registered 80°, and a few minutes after sunset (six p.m.) as low as 46° F. No food could be got within a very wide area, and if we were compelled to stay there, we would be starved. However, after a very large present had been extracted from us, and a very poor one given in return, we were graciously allowed to move further south to a large village where food was abundant.

The country we now traversed presented in its minor details some more variety than the country we had passed, for on leaving Mkubwasanya we had entered the district of Ubena.

There were still the same general features of undulating rounded ridge and valley, trending according to the line of drainage. The same dreary aspect prevailed; here a tract of scrub with an occasional large acacia-tree, euphorbia, or baobab, alternating with grazing-grounds, which at this period formed yellow fields of burnt up grass, making us wonder where the herds of cattle got sufficient sustenance to keep them alive. The soil over the whole of the country through which we had marched consisted of stiff red clay, which becomes in the dry season as hard as rock, and hardly supports any kind of vegetation. Hence the greater part is a sterile waste, incapable of affording grass for cattle, and useless for cultivation.

From the mountains to Mkubwasanya the underlying rocks are compact granite with occasional areas of volcanic rocks. After leaving Mkubwasanya the granite becomes more felspathic, and so extremely decomposable, that for a vast depth it has become transformed into a slightly sandy clay full of huge blocks of rock, which being more compact, have resisted longer the disintegrating agents. This clay has become cut into narrow gulleys from fifty to sixty feet deep. The effect is highly picturesque from the pink colour of the clay, the mud pillars which line the gorges and the shrubs and creepers that cover them. But besides getting cut into gulleys the general surface had been gradually worn away till the whole country seemed covered with colossal blocks. Some formed caps to pillars; some had almost reached the condition of rocking-stones, and everywhere they lay scattered in wild confusion, naturally making the unscientific mind conclude that here in this chaos is the evidence of some grand convulsion of nature, which has smashed the underlying rocks and hurled the fragments into the position they now occupy. But this would be a mistake. A moment's reflection shows that here we have only another instance of the "long results of time," the infinitely little working unceasingly in the falling rain, the changing temperature, and the corroding influence of carbonic acid.

The strange phenomena here presented have not been without their effect on the native mind. Instead of earthquakes and convulsions, of which he knows nothing, the savage sees only the work of demons and evil beings; and as he passes the place he tries to propitiate them by throwing a stone or a piece of wood into the numerous cairns which line the pathway.

This was the first time I had noticed this form of fetich worship, though frequently afterwards it came under my observation. Whenever anything strange and unusual in nature is seen by the native, he at once detects a ghost or a demon in it, with power to kill or smite with disease, but which can be appeased by some offering or other. Usually this occurs at the boundary of two districts, or where some dangerous tract of country commences. The common custom then is to throw down on a heap a stone, a rag, stick, or some grass; and so thoroughly do they believe in this practice that no one presumes to pass without such a tribute. Otherwise they would get killed or die from disease. Curiously enough, tribes who offer no such tribute to demons in their own country, carefully make the necessary offering, if they should pass one of these sacred spots beyond their own boundaries. The Waswahili are an example.

The dreadful east wind which had troubled us so long now began to veer to the south, and became daily more bearable and comfortable. Food also became more abundant, and we were the more encouraged in consequence.

On the third day we reached the village of Uhengè, where we were to await the pleasure of Mamlè the chief. Our place of detention proved to be pleasant, far beyond anything we had yet experienced in Uhehe. Food was abundant and varied in kind; and the men could enjoy themselves and recruit after their recent hardships.

From this place I determined to send three of my men back with letters to the coast. I directed them to return by way of Ujiji. If I had not arrived there as soon as themselves they were to cross the Lake to the west side and march south and seek for me. To the work of writing letters and reports I therefore set myself, and the time of our detention, which otherwise might have been irksome, passed agreeably.

We may now profitably take a rapid glace at the history of the Wahehe. We have seen in the case of the Mahenge how, from a very trivial circumstance, a cowardly and insignificant body of natives suddenly rose to great influence and power, acquiring a character for bravery which did not belong to them. In the case of the Wahehe we have another instance of the sudden rise into power and importance of a very insignificant tribe, but without the explanatory circumstances which, as we have seen, accounted for the rise of the Mahenge.

Till the time when the East Central African Expedition visited Uhehe comparatively little was known about the country or its people. Burton, in the journey already alluded to, passed a small corner of the country, and formed a far from favourable opinion regarding them. He describes the Wahehe as occupying a small area on the banks of the Ruaha, bounded on the north by Ugogo and Usagara, and on the west by Urori. "They are a plain race, but stout and hearty. They are decided pilferers. They carried off a flock of goats; and at night no man, unless encamped in a strong kraal, was safe from their attempts to snatch his goods. They are on bad terms with all their neighbours." They are further described as fearing to meet the Warori in the field. Such is the amount of information acquired by Burton regarding the position and character of the Wahehe.

Nothing more is learned regarding them till about two years before the Society's Expedition started, when we learn that Captain Elton and party were besieged some distance north of Lake Nyassa, in the stockade of Merere, chief of the Warori, by a tribe called Machinga, who were, however, repulsed. I subsequently learned that the Machinga were, in reality, the Wahehe, and had acquired that name from their chief.

After diligent inquiries, I have been able to form a pretty accurate notion of the later history of the Wahehe. It is as follows:—

Some ten or twelve years ago, while the Wahehe were still confined to their restricted tract of country between Ugogo and the Ruaha—safe among their high mountains after any cattle raid on neighbouring tribes—a chief named Machinga, of unusual bravery and energy, arose among them. Dissatisfied with his small chieftainship, and burning for extended power and wealth, he determined to wrest the country of Urori from the hands of the great and powerful chief Merere. He had acquired despotic power among his subjects, and they were ready to follow wherever he led. Thus commanded, the Wahehe crossed their southern border-line, and attacked the Warori or Wasango. In every engagement they were victorious. Villages were burned and cattle seized, and like an irresistible wave they swept across the plateau, devastating the entire country. Merere, unable to defend his immense town, burned it to prevent it falling into the hands of the Wahehe, drove off his princely herds of cattle, and took refuge in the higher mountainous plateau which surrounds the north end of Lake Nyassa. He here formed a strong stockade by the side of the Ruaha, and bade defiance to the hitherto victorious Wahehe. At this juncture Captain Elton and party on their way from Nyassa to the coast reached the seat of war, and were smuggled secretly into the stockade, where they found the besieged in sore straits for want of food. Their presence and their guns, however, gave Merere renewed courage, and finally the Wahehe retired disorganized and broken, like some great wave which had dashed against the solid rock.

Immediately after this defeat an under-chief called Mamlè formed a conspiracy against Machinga, encouraged doubtless by the reverse which the chief had sustained. Being sure of a section of the tribe, he murdered Machinga and seized the chieftainship, informing the latter's son that he was but a boy and not able to wield his father's sword. A part of the tribe, still faithful to the rightful heir, but too weak to resist, decamped during the night, and returned to their original country, north of the Ruaha where they again established themselves under Machinga's son.

Mamlè, now chief of the more important section of the tribe, once more took up arms against Merere, and this time more successfully. In every engagement the latter was beaten, till at last, unable to keep a footing in his own country, the great chief of all Urori, Usango, and Ubena, was compelled to flee into Usafa; and Mamlè reigned supreme from Mpwapwa and Ugogo to Lake Nyassa. It seemed like a fight between some small terrier and a huge bull-dog; but yet under the influence of one of those extraordinary impulses which will take possession of a people, under some great leader, this small band of Wahehe had been enabled to wreck apparently irretrievably the power of Urori.

In this position we found the Wahehe when we passed through their country in 1879. But for the continual appearance of newly-built villages, and the scarcity of food, there was nothing to mark the fact that we were in a newly-conquered country. They might have been settled there for centuries as far as outward evidence could show, and to all appearance the formerly insignificant tribe had been transformed in a few years into one of unusual power and strength.

But this state of things was short-lived: and though somewhat anticipating the course of events I may at once finish what remains of the history of the Wahehe. Even as I passed through the country influences were at work which brought about the downthrow of Mamlè's power. The son of Machinga determined to be revenged on Mamlè, and anxious to regain his rightful place at the head of the tribe, sent ambassadors to Merere in Usafa offering to join him in an attack upon Mamlè. Merere, with equal ambition to regain his place and country, at once accepted the offer, and great preparations were made to resume the war.

Four months after I passed through the country Merere was again in the field. On my return to the south end of Lake Tanganyika, after my exploration of its western side and of the Lukuga, I had intended to march for the coast straight east to Kilwa, by way of Usafa, Ubena, and the Rufiji valley. But unfortunately I then learned that the war was going on, and that it would be absolutely impossible to pass. In these circumstances I was compelled to turn northward. When I was in Fipa news arrived of the progress of the war. Merere was everywhere successful, and had already regained a large part of his country, including his large town, which he had commenced rebuilding. No further news reached us till on the borders of Ugogo, when we heard that Mamlè had been driven out of the entire country he so lately possessed, and, being unable to enter Uhehe from the hostility of Machinga's son, he had been driven with his few remaining warriors into Kiwere, where he had joined the robber-bands of Nyungu, a chief who has become notorious as the murderer of the missionary Penrose. The very village at which we received this information had been attacked by one of these bands, who were repulsed, leaving several dead on the field, a few days before I arrived; and I secured one of the fine shields used by the Wahehe, which had been taken from a slain warrior, and which is now safely housed in England. Of all the many fine villages which I saw in my march to Nyassa, there is probably not one now standing. Such is the evanescent nature of governments, peoples, and villages in Africa.

The Wahehe are upon the whole a rather good-looking class of negroes, not very dark coloured, and having very fine muscular figures. One of their most marked characteristics is the total absence of hair on their faces and bodies, and the scantiness of it on their heads. I never observed a single individual with the slightest suspicion of a beard or moustache. I have already spoken about their dress. In war they carry an elliptical fly-shaped shield of hide, about three and a half feet long, and eighteen inches broad at the middle. Their arms consist of a number of assegais, a stabbing-spear, and a hybrid article, between a billhook and an axe; they do not use the knobkerry, or club. They have great powers of endurance, whether it be in resisting fatigue or hunger. When occasion requires, they will travel at a trot for several days without food, thus being able to strike a blow suddenly and unexpectedly at an enemy. As a purely pastoral race they depend almost entirely on their cattle for food, and the men never condescend to work in the field. They, however, milk the cows, probably to prevent the wives from helping themselves to the produce, which they drink warm, the residue being left till it curdles, when it is much prized and said to be ripe. When flesh meat is eaten, it is simply warmed on the surface and then devoured. They have neither goats nor fowls. The absence of the latter is very remarkable, as in every other part of the country I have visited I have found them abundant. The climate can hardly explain it, for in Ugogo further north, with much the same conditions, they are found in great numbers. The war is probably to blame, as in moving from place to place they could not be encumbered with them, and hence they would be killed and eaten.

The Wahehe have almost no domestic utensils. One or two calabashes, an earthenware cooking-pot, and a wooden vessel comprise all their furniture, with the exception of the ox-hide on which they sleep. When migrating and on the march they are thus free from all encumbrances, as the cattle can easily be driven. The wooden vessels into which they milk the cows are always very filthy, having a thick coating of dung; in most other respects they seemed more cleanly.

I am inclined to estimate the character of the Wahehe at a much higher standard than Burton. They may have been decided pilferers when they lived more as a cattle-lifting band on the southern confines of Ugogo, but their sudden rise to power and their greater acquaintance with traders and experience of the advantages of trade, may have raised their morality to a higher level. This I can say of my personal experience among them: I have never had a single article stolen from me or from the men. Frequently, we were all scattered over the country, seeking food and leaving only a few men to guard the camp, which was frequently on such occasions thronged with Wahehe, yet no one put forth his hand to touch what did not belong to him.

There was a certain style and dignity in the movements of the chiefs that faintly reminded one of that pet of some novelists, the noble Indian of the forest or prairie. They rarely were openly rude, and anything disagreeable was as a rule gently suggested, leaving us to infer what was meant. Thus in the matter of hongo or tribute, they always spoke of it as an interchange of presents between gentlemen. They never were so rude as to tell you they were not satisfied with your present and wanted more. Instead, as in an instance already noted, they discovered to their infinite regret that their cows had stopped giving milk, or that the times had been very hard and no food was to be got, or worse still, in spite of their power all the guides had suddenly become obstinate and could not be advised to conduct us to the next village. An addition to the present usually wrought startling miracles. The cows became lavish of their milk, and the very dirt seemed to turn into corn to fill the granaries of Uhehe. A further addition would even thaw the hard hearts of the guides, and draw forth expressions of wonder at the generosity and open-handedness of the white man. We were thus gently eased of our bales to an alarming extent. But then we felt that among such noblemen of Nature's mould, it would be sheer niggardliness to be less lavish. We could not but spend our wealth handsomely.

The Wahehe appear to be as free from superstitious notions as any tribe I have seen. I never saw one Mganga or medicine-man. Charms were rarely worn, and indeed there was an entire absence of the usual signs of that fetichism, which is so prevalent elsewhere. They seem, however, to have no respect for their dead; the bodies being generally thrown into the jungle to be eaten by the hyenas which prowl about during the night, filling the air with their vicious snarls. This seeming want of respect for their dead may be accounted for by their warlike and migratory habits. The more settled and peaceable a tribe is, the more superstitious it is.

It is seldom, however, that the geographical explorer can form just conceptions of the manners or customs of a native tribe, or of their moral character, travelling as he generally does straight through the country, and meeting the natives but a short time, when under the influence of fear or suspicion of the great man. It can only be by a prolonged residence in the district and a thorough command of the language that a person is entitled to speak with the confidence of certain knowledge. The Central Africans have not had this justice done to them, and till such justice is done, we have little right to draw very definite conclusions about the negro mind. To me it seems that most travellers under the influence of fevers and the thousand troubles attendant on African travelling, have much maligned and unjustly abused the natives, and that few people have studied them with unprejudiced and unbiassed minds.

We have already noticed the main features or characteristics of the Plateau we had crossed at a general elevation of 4000 to 5000 feet. Over these barren, clayey ridges and hollows little of life is seen, if we leave out of consideration the herds of cattle. The sight of a few quagga, or zebras, was usually sufficient to raise a general excitement among the men. A few antelopes were twice descried, and I heard of an elephant having been killed at a place where we encamped. This exhausts the amount of game seen or heard of over a stretch of upwards of 200 miles of country. At night numerous hyenas prowled about our camp, and several of the croenta species were shot. The country is too cold and bleak, and there is too little game for the lion, which is never found in that region, and the forest-loving leopard is equally absent. A species of jackal was on one occasion seen.

The absence of the feathered creation is equally marked. A few crows about the rocks, or vultures, ugly and repulsive, watching for some carrion, are the only prominent birds. Very rarely does another show itself.

Curiously enough there is more variety of flowers in these wastes than in the richer lowlands. In the tropics, where everything is favourable to a luxuriant vegetable growth, Nature usually spends her energies in producing an infinite variety in the form and green tinting of the foliage. She delights in grand colossal trees, up the trunks of which grow strange creepers, and from whose branches they hang in pendent masses of graceful foliage. She rejoices in tender, beautiful ferns, or in majestic palms, but rarely in brilliant flowers. Fine flowers are scarcely even an appreciable element in tropical vegetation.

Mr. Wallace, in his "Malay Archipelago," has well remarked "that the bright colours of flowers have a much greater influence on the general aspect of nature in temperate than in tropical climates." He further adds, "During twelve years spent amidst the grandest tropical vegetation I have seen nothing comparable to the effect produced on our landscapes by gorse, broom, heather, wild hyacinths, hawthorn, purple orchises, and butter-cups." These words descriptive of the Malay Archipelago are equally applicable to Africa. On the damp lowlands, as in the midst of the grand forest scenery of Usambara the same fact was equally noticeable. Fine and varied foliage everywhere attracted the eye, but gay flowers never. In those parts, however, which more nearly approximate to the temperate condition of our own country, such as in the great Central Plateau, we have an approach towards the same prominence in the flowering plants. It was a great delight to me in my many weary tramps to have my thoughts drawn away from brooding on my troubles by the sight of the numerous pretty little flowers which everywhere had sprung up after the burning off of the grass. They all looked so familiar and homelike that I experienced the greatest enjoyment, plucking them here and there for my collection. Among trees and shrubs the only prominent forms were proteas, which frequently covered considerable tracts, that vegetable monstrosity the baobab, or its equally eccentric neighbour the tree-euphorbia. Occasionally an acacia or mimosa added to the variety.

With these general notions regarding the Wahehe and their country we may now return to Uhenge.

Our appearance at this important place was made the occasion of public rejoicing; for the natives are ever ready to seize any opportunity of getting up a fête. They drove me almost mad dancing in hundreds about my hut, thumping innumerable drums of all sizes, notes, and tones, clapping hands, and singing most lustily, till I rolled about vociferating inwardly in my despair, and wondering if after all I had got unawares into some minor pandemonium. If tears would have been of any use I think I should have humbly dropped a few to appeal to their compassion; but as I reflected that they might put a wrong interpretation on them I refrained and prayed for patience. The hard clay about the villages is usually tramped by the cattle into exceedingly fine dust, which blows in continuous clouds during the day, till one feels perfectly choked. In ordinary circumstances that was bad enough, but when a few hundreds danced in it then it became almost necessary to wrap a blanket round my head, and lie down patiently as in an African sirocco.

After I had finished my letters and reports, and despatched them by three of my men to the coast, I had to try various means to occupy my time, as there was still no news of Mamlè, and how long my "honourable" imprisonment was likely to last, was a question of uncertainty. I had several amusing interviews with the village chief, to whom I showed my various strange instruments and curiosities. With children's fancy pictures of natural history objects he was greatly taken, more especially whenever he saw anything he recognized. But my great *coup* was the exhibition of the portraits I had with me. At first he examined them with a meaningless look as if not quite understanding what they were, but when he was told that they were the pictures of white men and women his rapture and astonishment knew no bounds. Their dresses, however, he could not quite comprehend, and was inclined to believe they were parts of the person. One photograph of a lady had a piece of lace round the shoulders. This he confidently asserted must be tatooing. He flattered my vanity very much, or rather Chuma did, by pointing out all the pleasant-looking young ladies as my wives. I must confess that I did not deny the "soft impeachment," for which I now offer my humble apologies. The chief under this misapprehension, and imagining them to be at my command to sell or dispose of as I pleased, wanted at once to strike a bargain, and have a consignment of these fair ladies sent to Uhenge.

As the chief rejoiced very much in hearing shooting, we had great fun caricaturing the evolutions and movements of his Highness Syed Bargash's troops. Sirkari, a freed slave from the palace, took command of the porters, and made some ludicrous attempts to make them march and turn in order, to fire in volleys, or regularly along the line. We thus contrived to put in the time for five days, when at last Mamlè, having satisfied himself that we were harmless, and had no evil intentions, made a sudden appearance at Uhenge. The more immediate reason of his coming, however, was the expectation of receiving an addition to the already large present we had given him. He was a coarse, bony savage, with an unusual amount of hair on his face. In the matter of dress he was like the rest of his subjects. At our interview, he informed me that he did not want to hide from us, but that he was curious to see what I would do. He was very much surprised at the way I hurried through the country, stopping never more than a day anywhere. How could I go back and tell my people that I had seen Uhehe and its people if I went in this way? His hand, he said, was against all his neighbours, and he did his utmost to carry off their cattle and women; but he did not fight with white men and Arabs, as he would then not get beads and cloth. He now wanted to make brothers with me, so that I and all my men would be safe wherever they might be found; otherwise they might be killed.

Next day at noon, preceded by our band and escorted by our chief riflemen, we went in great state to the chief's house, where the important ceremony was to be performed. On both sides proxies were chosen,—Chuma for me, and the headman of Uhenge for Mamlè. The proxies sat down opposite each other with legs interlaced. The spear of the one and the gun of the other were placed across their respective shoulders. An Mhehe then made three small cuts in the breast of Chuma, and Uledi, my gun-bearer, made the same on the chief's breast, at which he visibly winced. Thereafter another Mhehe took the axe of Mamlè's proxy and rubbed it on the spear, making a speech the while, drawing down curses on the latter if he should "break the brotherhood" by doing the other harm in any way, and swearing that he shall stand by his brother whenever danger threatened. He invoked the gods (whatever they may be) to turn the weapon of the traitor against himself, and that he might have no peace or welfare if he did not religiously adhere to all the requirements of the brotherhood. Next Uledi took Chuma's knife and rubbed it on his gun, and made a similar speech. The ceremony was finished by a small piece of meat being taken and rubbed in the dripping blood of Chuma, and given to the headman to eat, the same being done for Chuma.

On the following day we gladly renewed our march towards Nyassa. In the afternoon, while camped at the tembe of Kahanga, and sitting outside my tent in the cool shade, I was suddenly surprised by the appearance of a troop of women dancing. Now if there is one thing on this earth

I abhor it is a dance by almost nude savages. One can sit through a good ballet with wonderful equanimity; and the whole, as seen on the stage, has such an air of unreality about it, that we are composed, and do not remark the predominance of legs and the scantiness of skirts. But in savage Africa it is different. There is an air of intense reality in the entire performance, and reality of such a low type as makes one shudder. There are no skin-tights or glittering dresses, and everything speaks of a condition but little removed from the brutish. Quite unaware of the sensitiveness of my feelings, these dark damsels surrounded me, and commenced their dreadful bacchanalian exhibition. They flung their arms and legs wildly about, they screeched till their voices cracked, and the more I fidgetted about, the worse they became. I looked at the skies, and I looked at the ground. I turned my back, and I stuffed my fingers in my ears. I yelled at my men to turn them away; but they wisely kept out of sight, and enjoyed the whole performance unseen. At last, unable to endure it longer, I made a wild rush and gained my tent. I crammed my head into the blankets; but in vain, for louder grew the song round my tent, till I threw out a considerable piece of cotton, which the hags clutched and bore off in triumph, leaving me almost exhausted by my impotent rage.

After leaving Uhenge there was little of interest to mark the progress of our movements till the 12th of September, when we reached the base of what appeared an extremely precipitous mountain, trending N.N.W. and S.S.E., and rising from 3000 to 4000 feet above the general level of the Plateau. We hailed this sight with pleasure, as a sign of our approach to the magnificent range of mountains described by Captain Elton, and supposed to be from 10,000 to 12,000 feet in height. The abrupt rise of these mountains, together with the evident signs of some ancient disturbance, seems to point to a line of fault by which the rocks to the south have been upheaved, or those north of the line lowered. At the base of the mountain a large stream, called the Mbangala, flows north and west to join the Ruaha, near Utengula, Merere's big town. The sections exposed in its course reveal frequent outbursts of ancient eruptive rocks in dykes and veins through the granite and metamorphic rocks, which they have broken and altered to such an extent as to put them almost beyond identification.

The people about this place are not Wahehe, but Wabena, a sub-tribe of Warori, of a servile character, who submit themselves without a murmur to whosoever happens to be master of the field. Cloth being scarce and dear, the men go absolutely naked, while the women have only a few strings of beads. They are by no means so good-looking as the Wahehe.

Following up the course of the Mbangala, we rose considerably in altitude, leaving Ubena behind us, and entered another tributary tribe, called the

Wapangwa. These were even more ugly than the Wabena, being exceedingly dark in colour, and having in them a good deal of the typical negro. They were nearly all, however, specially characterized by a cast in the eye, which was so marked that they well merit the name of the squinting negroes. This must be owing, I think, not only to the hard life they lead, but also to their birth, as other tribes living under as bad conditions are not characterized by this defect. Besides squinting, a very large number are blind in the left eye. Why it should be always the left eye I have no means of ascertaining. Here is a problem for the advocates of "inherited variations." How did the desire to squint arise, and how did such an obvious optical defect come to be perpetuated in succeeding generations? How also is the left eye in almost every case the one which becomes blind among the Wapangwa? Probably this latter may be some form of punishment inflicted by the chiefs. I can account for it in no other way; but unfortunately having made no inquiries whilst among the people, I can do nothing to prove the truth of this theory now. Still these facts remain, that among the small tribe of Wapangwa, living at a height of 6000 to 7000 feet among treeless mountains, the majority of the men squint, and a very large number seem as if they had had the ball of the left eye quite destroyed.

At the village of Msendisi, where we camped, the chief whom we interviewed was too much clothed in his dignity to speak, though during the visit his teeth quite chattered in his head with cold. As in Mahenge our discourse had to go through many mouths, I spoke to Chuma, Chuma to our interpreter, he to the guide, the guide to a councillor and he finally to the chief. The answer then commenced from the councillor, and came back to me in reverse order.

Though we had arrived so near to the lake, yet we never could find any one who had ever visited it. They all betrayed the most extraordinary ignorance of the geography of the country beyond a few miles, and I never could get guides to go beyond the nearest village. All the way through Uhehe and Ubena the people had no known route to the lake, so that I had always to point in the direction I wanted to go, and ask to be taken as near it as possible. We were thus continually at the mercy of the chief, who made it the means of the most vile extortion (though always in the most polite way). Day after day we had to give large presents to get guides.

Leaving Msendisi we crossed a ridge to the Mbangala, then up another very steep one, at the top of which our guide pointed out to us the Ruaha, where it actually commenced as such. From our vantage-ground we could see the three small rills, by the joining of which the Ruaha is formed, here only a small stream. Our route had thus crossed it at its starting-point, as formerly it had crossed it nearly at the point where it joins the Uranga and becomes lost in the Rufiji river. From the ridge we could trace it for some distance running north-west through a deep narrow gorge.

The surrounding scenery, though wanting in variety, was yet very fine—the almost perpendicular grassy mountain sides, cut here and there into gullies, where springs trickled down, and filled with a deep green vegetation, contrasting well with the yellow grass which covered the rest of the mountains. Not a tree was to be seen.

After crossing the Ruaha gorge, and reaching the mountain top on the opposite side, we camped at the village of Mwangwama, inhabited by another small tribe called the Wanena.

If the Wapangwa looked of an unusually low type the Wanena were certainly infinitely lower. Here, indeed, were sufficient marks of degradation to arouse the enthusiasm and scientific delight of ethnologists and those who seek after connecting links between apes and men. Such miserable specimens of humanity I have nowhere seen in Africa. Even the Wakhutu looked intelligent and manly beside them. Their heads are very small, and their skulls are so narrow as to suggest the idea that they have been pressed so when young. The two upper incisors are remarkably large and prominent. Their language is different from any of the surrounding tribes, and we found it quite impossible to communicate with them except by signs. The men wear a small piece of hide over the shoulder, and the women's dress is reduced to a bunch of grass. They live in the most wretched hovels, beside which a hole in the ground would be comfortable. These hovels are conical in form, eight feet in height, and seven feet in diameter. They are built of a few inclined sticks tied together at the top, and then rudely thatched with grass, leaving a hole eighteen inches high to serve as a doorway. My men would have been ashamed to have run up such a miserable structure in half an hour, and yet whole families huddle themselves like so many pigs in them, and must necessarily lie on each other. I observed no domestic utensils, though doubtless they have, at least, an earthenware cooking-pot. Their chief food is peas, at least that was the only kind of diet we could get. Of these they eagerly brought basketfuls to barter for a strip of cotton an inch or two inches broad. They are exceedingly timid.

It is somewhat strange that while in almost every other part of the world the mountain tribes are the bravest and noblest, in Africa it is quite the reverse. Whenever inhabitants are found occupying mountains in Africa, they are in almost every case some miserable little remnant of a tribe, few in numbers, cowardly in the extreme, degraded in mind and body, and living worse than the brutes; never sure of their lives or of food for the morrow; everything, in fact, speaking eloquently of the very dregs and depths of misery. This may be accounted for on the supposition that no African will take to the mountains till he is compelled to do so to save his life. In the lowlands nature supplies him abundantly with food in return for the minimum of work. The climate is warm and equable, and necessitates no clothing. On the mountains it is different. There he must work hard for his living; he is subjected to the greatest extremes of heat and cold, and all his surroundings are wretched. Hence no tribe will ever betake itself thither till on the verge of annihilation; and then it retires, broken up and disorganized, with no unity and no spirit. Thus during war-raids the only safety is to retire to still greater fastnesses, and to still more miserable conditions of life. Communication with the outer world becomes cut off; lower and lower it sinks mentally and physically, till such tribes (if a few score of people can constitute a tribe) as the Wanena, Wakinga, and Wapangwa are formed. I am confidently of the opinion that these tribes represent a degradation from a formerly higher level. Not unlikely they have descended from the former possessors of the continent, who have been gradually overwhelmed by the rise of new and perhaps more energetic tribes.

For three days we continued our progress through mountainous country at a general level of 7000 feet, and still we could observe no range answering to the grand Konde Mountains of Elton. Nothing but irregular hummocky masses of rounded mountains met our eye, trending in various directions, according to the drainage. On leaving Mwangwama, a village of Wanena, we crossed the watershed between the waters which drain into the Ruaha and those which turn south to Lake Nyassa. It was clear then that these latter must cut through the Konde range, which made me doubt altogether the existence of such a range. The natives, moreover, knew nothing about it.

At last, on the 18th of September, we stood on the country where the range should have been, but I looked in vain for it; the same general level prevailed away to the far horizon. We camped at an Ukinga village named Mtandala, and the day being a Mohammedan feast (or as Chuma called it, their Christmas), I gave all the men a present and a holiday. However, I myself could not rest, but started off to what seemed the highest mountain in the neighbourhood. After a hard climb I reached the top, and found it to be only 8000 feet in height, and as my guides pointed out where Nyassa could be seen on clear days, I was reluctantly compelled to conclude that though there might be a Konde mountain, there was certainly no Konde range. I further concluded that there was no range at all round Nyassa, and that what seemed to be such, viewed from the lake, was only the eastern escarpment of a plateau, varying in height from 6000 to 8000 feet above the sea, encircling the north end of Lake Nyassa, and extending away westwards round Lake Leopold,—its western escarpment forming the Lambalamfipa Mountains in Fipa. This plateau consists of clay, slates, and other metamorphic rocks, except, as we shall see presently, around the lake, where the rocks become volcanic. The general appearance is very monotonous; it is characterized by smooth grassy heights, rounded and unbroken on their tops, but having steep sides. There are no trees, no serrated ridges, no precipices or rocky picturesque scenes. Such is the description of the greater part of this tract of country, which well merits the title of the Upper Plateau of the Lake Region.

When we reached Mtandala we were almost within sight of the lake, towards which we had so long and wearily tramped. Yet I felt as if it required more strength than I possessed to reach it. The frightful fever which attacked me on first leaving the lowlands, still clung to me like a vampire, sucking my heart's blood. But for the rule I adopted I think I should never have survived to get so far. That rule was simply to keep marching on as long as my legs would sustain me, and never to be carried by my men. For an African traveller to halt that he may get better is the worst policy possible. With nothing to amuse him, he lies and groans in his tent. There is little to think of but his troubles, his delays, and his thousand and one vexations, while a burning sun pours down on his flimsy tent.

He is thus apt to become worse rather than better. But when on the march it is different. He has his route to think about; he has to note the characteristics of the country, to hold consultations with guides and chiefs, and keep a vigilant eye on the men, so that everything may be straight. His mind is thus drawn away from his troubles. The physical exertion helps to keep his system in better working order, and enables him to eat and sleep.

On this principle then I moved on. I never allowed my illness to stop me a single day. Perhaps some may be sceptical when I say that I have frequently marched till I have fallen on the road, but carried I would not be. Poor Johnston's tortures rose too vividly before my eyes. For more than a month at this time my memory entirely forsook me, so that I have sat at night vainly attempting to write up my diary, and almost tearing my hair in despair. The day's journey would appear as a perfect blank, and I would sit and stare at vacancy like an idiot, when I tried to think whether or not we had crossed any streams, passed any villages, or seen any mountains. I could not remember the names of any of my friends. And thus I would sit in the most ludicrous plight, endeavouring to recall what had marked our course. I have actually a clearer idea now of the incidents of each day's march than I had at the close of any particular day, and but for my constant habit of jotting down at the moment whatever occurred or was seen, my map and diary would have shown a considerable hiatus. And now when I had got almost within reach of Nyassa I was so weak that I could hardly lift a weight of six pounds. Under these conditions mountains were crossed, and more lay before us.

On leaving Mtandala we had a long march to reach the next village. The men got into camp after midday, but I was so far behind with my good servant Uledi that the men became alarmed, and came back with a hammock to carry me. That I knew would only have been torture amongst such precipitous hills. However, as I could not get along, a rope was tied round my waist, and with this I was hauled up the mountain, while two men steadied me behind. I was thus brought triumphantly into camp like a valuable prisoner of war.

The scenery around Paparua, the village where we camped, was of a completely different type from anything we had yet seen. The landscape was grand and striking in the extreme. There were sharp jagged peaks, vertical rocky sides, notched and cut in the most irregular fashion. To these were added serrated mountain tops, yawning gorges, and great precipices, where vegetation in vain attempts to grow. Arrange such features in your mind and some notion of the scenery will be obtained.

Next day, feeling somewhat better and invigorated by the news that we should see Nyassa, if not actually reach it, that day, I started with the rising sun. Rounding the shoulder of the hill, we crossed a wooded glen, and climbed to the top of a prominent ridge. With a hearty shout the men announced the sight of the lake. With dazed eyes I looked in vain for some time. At last through a gap in the serrated mountain I descried a blue space, which gradually separated itself from the azure of the sky, and revealed itself beyond a doubt as the lake. I had no words to express my joy, but sat down and gazed earnestly at the prize I had striven for.

We were soon celebrating the event in another manner. The porters, knowing how to get round me, had quietly set all the natives at work over night to brew pombe; and now these came at the proper moment all in a string, and seemingly passing by, but offered to sell it. Of course I could not deny the men this indulgence. So the luxury was bought, and every man drank his fill, with appropriate speeches.

We were all too much exhausted to get very far that day, so we camped at a small village at the bottom of a deep gorge. I was too restless, however, to keep still, and in the afternoon could not resist the great longing to climb the nearest mountain alone, to feast my eyes once more on Nyassa. This I accordingly did, reaching the top just before the sun set, lighting up the sky, the waters, and the mountains with the most gorgeous golden hues. The northern outlines of the lake could be clearly discerned; the magnificent plateau escarpment to the east, known as the Livingstone Mountains; and the rich level plain of the Jumbaka to the west, with its silvery winding streams and subtending plateau.

Between me and the lake lay two gorges, three or four thousand feet deep, divided by two precipitous mountains. These had still to be crossed before the lake was reached, and I looked at them with dismay. The sun set, the twilight passed, the moon rose, and still I sat and dreamed, till I was rudely awakened to my situation by a deep sullen roar from a neighbouring hill. I jumped up in affright, for no one could mistake the voice of the lion. It was an exciting position to be in—two hours from camp, at night, on an uninhabited mountain, without a gun. With nervous alacrity I took to my heels, slipping and tripping, but heeding not; for again came the deep roar of the lion. Leopards and hyenas also infested the mountains, and I felt I was running the gauntlet among them as my excited imagination turned every rock into an animal. Half way down the hill I was met by my men, who, alarmed at my long absence, had turned out *en masse* to search for me. I nearly fainted (owing to my illness) when I reached them, and had to be helped into camp.

Next day we got on the march, with cries of "Nyassa! Nyassa!" Mounting to the top of the mountain I had visited last night, we descended with much difficulty the opposite side into the deep gorge. Then up 3000 feet we rose again to pass the top and descend once more.

This second descent was one of much danger and difficulty. A false step, and a dozen men would have been hurled down hundreds of feet. If a load had fallen it would have swept every man before it. Every possible precaution had to be observed not to loosen the rocks. The loads had first to be laid down; then the porters descended a few steps, and lifted the loads down to a lower place; and so on till the bottom was reached.

We rested two hours beside the stream which had formed this gloomy gorge, to allow stragglers to come up. We then commenced the ascent of the last mountain. In this case the loads had to be lifted up so far first; then the porters had to scramble up beside them with hands and knees, each one assisting his neighbour. It soon became evident that Nyassa could not be reached that day, as the sun had almost set, and every man was dead beat. So we camped on the hill side.

We here discovered that one of the men was missing; but only by the offer of a high reward could I get any one to go back and search for him. No one of the searching party returned that night. They must have slept on the bare mountain side, without food, after their hard day's work.

Next morning they did not appear. We waited a few hours in vain. Nyassa, however, must be reached that day; so I sent ten men back, with hammocks and nourishing food, to look for the lost party, and then we started for the lake. I was at first in the rear, but gradually my excitement got the better of my weakness, and I pushed past one man after another till I got to the front. At last the mountain top was reached, and there was Nyassa at our feet, 3000 feet below us! With a wild halloo I now rushed on. Down the hillside I went at headlong speed, in momentary danger of a capsize, with probably fatal consequences.

An hour later, and we hear the faint roar of the Nyassa breakers, wafted by a cool breeze to our ears. Another hour, and a few of us go rushing like madmen into a village at the bottom of the mountain, sending the natives flying in terror. We reach the side of the lake. Off go our clothes; and as I dash into the water, I feel myself baptized an African traveller.

CHAPTER VII.

BETWEEN NYASSA AND TANGANYIKA.

The point where we reached the shores of Lake Nyassa proved to be almost at its most northern extremity, and I may fairly lay claim to the honour of being the first who ever reached it. It is true that both the Livingstonia missionaries and the party under Captain Elton had come within seven or eight miles of the same place; but they had never got to the actual termination of the lake. We pitched our camp in the village of Pupanganda, under the shade of some magnificent fig-trees. The village occupies a small piece of ground gained from the lake by the *débris* brought down from the hills during the rainy season. The precipitous mountains rise in picturesque peaks and serrated masses, clothed towards the lake with forest, and over these come tumbling many noisy cascades. Except on the little strips of ground occupied by Pupanganda, the mountains tower up in almost sheer precipices from the water's edge, not even leaving sufficiently good footing to form a pathway. Night and day, in storm or in calm, breakers constantly roar as they dash themselves against the almost adamantine rocks. But how glorious it was to us to lie in the cool shade and listen dreamily to the soothing sound, or be sent at night asleep by the continuous cadence of the waves. We felt as if our work was over, and that we had reached a promised land.

But unfortunately we had still cause for alarm in the non-appearance of the missing men and of those who had been sent from Paupwe to seek for them. Although nearly every man was half dead with fatigue when we reached the lake, it became evident to all that more help must be sent off at once with food, or they would all be starved. A few volunteered like heroes to return in search of their comrades, and the necessary number was soon made up by impressing into the service a few of the least-heavily worked. The sun was setting when they commenced their march, and it was clear that they would have to spend at least one night on the hills. During the hours of darkness we watched anxiously for signals from the heights, but none were made. In the morning there were still no signs, and we concluded that the original absentees must have fallen over a precipice, or got devoured by wild beasts.

While we thus waited I strolled about the village to observe the inhabitants and their occupations. The men were a considerable improvement on the mountain tribes we had just left, and presented the appearance of average East African negroes. The women, however, were more prominent in their peculiarities. They had remarkably small heads, very long, swan-like necks, with a most abnormal abdominal development, tapering like a cone up to the neck. Their legs were long and thin, and the length of the mammæ was something altogether exceptional even in Africa. The left breast was generally longer than the right, owing to their habit of carrying their children on that side. One woman I noticed had her left breast about fifteen inches long, and she could with the utmost ease suckle her child over her shoulder. Altogether, the women looked such unattractive specimens as are rarely to be seen.

The huts of these people were of the Wakhutu form, but seldom so large, and having taller walls and smaller roofs in proportion. They have the mud between the poles kneaded, and shaped in the form of rounded bricks.

The villagers gain their livelihood entirely by fishing and pottery. On all sides the various stages in the making of pots could be observed, the digging of the material from a tuff bed in the mountain side, the working of it into plastic grey clay, the skilful and artistic manipulation of the clay into pots of all sizes and shapes, the final adding of smoothing and finishing touches, and the hardening by fire. It was a never-failing source of astonishment to me to observe how quickly, and yet with what geometrical exactness, they formed these pots. The women were the principal workers, and each one was able to make two or three in a day. They were chiefly pombe pots, and so large that one could be lifted only with difficulty. Pupanganda is famed all over the surrounding country for its potteries, the fine grey tuff being peculiarly well adapted for the purpose.

At night, much to our relief, the entire party from the hills arrived, bringing two of their number almost dead. But two days of beef-tea and quinine put them soon on their legs again, although they required to be forced like children to take the latter nauseous stuff.

I was much excited on my arrival at the lake to hear that there was a white man staying at Mbungo hunting, and I at once sent off a letter by Stamboul to inform him of my arrival. On his return I was puzzled by his strange report. In fact I could not make out anything beyond the fact that some one in European clothes was there, but evidently not an Englishman, as no answer was returned to the letter.

On the third day after our arrival, while I was lying idly passing away the time like some lotus-eater, Chuma disturbed my dreams by appearing before me dragging a reluctant native damsel after him. When he had somewhat recovered his breath he informed me that he had just discovered an unattached female in the camp, who it appeared had run away from Paparua, had clandestinely joined our caravan with the wonderful aspiration of reaching the coast, and had eluded his observation for several days. What was he to do with her? This seemed to be rather a difficult problem to solve. If we cast her out of the camp she would inevitably be made a slave of; and yet it would be demoralizing to have such a woman with us.

In the perplexing circumstances I ransacked my memory for a precedent among the experiences of travellers. Happily I remembered that Livingstone himself had been in the same perplexity about an unattached damsel, and had there and then married her to no less a person than Chuma himself. The matter then seemed plain. "We must get some one to marry her," I said to Chuma, whereat that worthy grinned, and looked askance at the lady. The hopes of such a consummation were certainly not great, as she was by no means a beauty—if she had been one, Chuma would have thought twice before troubling me with her. She was blind in the left eye, and the other had a decided tendency to look sideways, so that when she wanted to walk straight forward she required to twist her head round to bring the right eye to bear forward. When she wanted to see anything on her left, she needs must turn right about. Her figure was far from being attractive; but though somewhat loosely thrown together she bid fair to be of good standing material. I therefore determined to give her a chance, so the drum was beat to call the men together. Before they all assembled I gave her a yard of cotton to hide a few of the more glaring defects, and in this bridal dress she was offered in marriage to whosoever would have her. Chuma did his best to appraise her, pointing out her strength, and how generally useful she would be to any one connubially inclined. There had been a general burst of laughter at first. But at last Stamboul, who had always shown himself very fond of domestic comforts, was overpowered by the argument that a pair of good hands were in the long-run better than a pretty face, and much less liable to be run away with. So he boldly offered to marry her. This settled all difficulties at once, and in less than half-an-hour we married them right off. Thereafter followed the customary festivities. It may be added that Mrs. Stamboul stuck faithfully to the caravan till on our return march to the coast, when lagging somewhat behind one day she disappeared, and was no more heard of. Whether she deserted or was kidnapped we never knew, for, sad to relate, her husband never went back to seek her.

When all our men were once more together and somewhat recruited we began to think of resuming our march to Tanganyika. I was rather anxious about the reception we might expect from the chief Makula and his people, being mindful of the big hongo which Elton and the Livingstonia missionaries had to pay, and the numerous troubles they had experienced among them. Still I saw no way to avoid them, and resolved to put my trust in the natives, as I had hitherto done with much advantage.

On the 25th of September we started for Makula's. The way proved to be exceedingly difficult. There was no passage along the edge of the lake, so we were compelled to climb the savage mountains. We required to take every advantage from roots, creepers, and projecting rocks, to make the ascent, and at many places the porters had mutually to assist each other.

At last about midday we fairly rounded the corner of the lake, and reached a great marshy expanse of country bordering this part of the Nyassa, and forming the country of Makula.

At the point where we left the mountains we discovered a river called the Lufira, falling into the lake, and probably the same as the Ruingero, seen by Elton near Mazotè's. We required canoes to cross it.

In consequence of the marshes which here border the low-lying country, we were compelled to keep along shore in deep loose sand, which made it dreadfully hard work to get along. After crossing the mouth of the river Jumbaka, we all got lost among swamps, and so broken up that every man had to find his own way.

At four p.m. attended only by my servant, I reached the outskirts of the village, and not knowing where the rest of the men were, I boldly marched in. The scene that opened up before me I beheld with astonishment. It seemed a perfect Arcadia, about which idyllic poets have sung, though few have seen it realized. Imagine a magnificent grove of bananas, laden with bunches of fruit, each of which would form a man's load, growing on a perfectly level plain, from which all weeds, garbage, and things unsightly are carefully cleared away. Dotted here and there are a number of immense shady sycamores, with branches each almost as large as a separate tree. At every few spaces are charmingly neat circular huts, with conical roofs, and walls hanging out all round with the clay worked prettily into rounded bricks, and daubed symmetrically with spots. The grass thatching is also very neat. The *tout ensemble* renders these huts worthy of a place in any nobleman's garden. Not an unsightly object was to be seen anywhere Indeed, one could not but imagine that he had entered a garden rather than a village, as only two or three houses could be seen at any one place, peeping out from the dense tropical grove.

The people accorded in every respect with the scene. The time was the sultry afternoon hours. The work of the day was over, and all were having a siesta before bestirring themselves to enjoy their evening repast and the cool hours of sunset. They lay in the shade of the huts, or under the sycamores, spread out on banana leaves, some with their faces down and others with them up. They were utterly naked; and there they lay as God had made them, unconscious of any want, and apparently fearing no danger. A drowsy sultriness hung around, and not a sound was heard, except the occasional buzz of a bee or a beetle, the flutter of a bird, or the roar of the neighbouring breakers mellowed into a sleepy sound by distance. I felt as if I had fallen upon some enchanted place, as I stood unseen, taking in the picture.

The charm was easily broken. We simply gave a shout, and in a moment every man was on his feet, clutching his spear. They had, however, no sooner set eyes on me than the form of the enchantment seemed changed. They stood transfixed like statues with surprise and terror at the strange apparition. It was only by repeated cries of "Friends, friends," and "Mzungu," that they were gradually brought to their senses, and prevented from making a precipitate retreat. As soon as they thoroughly comprehended who we were, they gave free vent to their delight, and ran about dancing like madmen.

Our men dropped in by parties, until we had all got once more together; and then we enjoyed such a feast as we had not seen since we had left Uzaramo. Bananas, sweet potatoes, yams, cassava, ground-nuts, Indian corn, millet, beans, &c., together with abundance of capital milk and beef, came pouring into the camp, to be disposed of for almost nominal prices, so eager were these savages for the most trifling bits of cotton. Next morning we were informed that we would see the white man, and also Makula. Amidst such plenty we were quite content to wait, though somewhat surprised that the white man had not appeared at once.

We had to shift our camp next day to a more convenient place. On starting I was electrified by the sudden appearance of a military coat, the possessor of which took hold of my hand and shook it, with a broken "Good morning." Imagining this to be the white man's servant, I inquired where he was, when, to my intense disappointment, it turned out that he was at present at Livingstonia; and so a cherished hope was shattered.

After a short march of an hour through richly-cultivated gardens and groves of bananas, we camped under a huge sycamore, which gave shade to the whole camp.

We were not well settled down before the chief, Makula, arrived. I was at once favourably impressed by him. He was young, tall, and well made, with frank pleasing features. His dress was by no means regal, or even becoming, consisting entirely as it did of something less than half a cast-off European shirt, with the major part behind. Acting on my advice, he altered this, with much advantage to European eyes. The only other articles worn were a number of iron bells round his ankles, which he made the most of by stepping out with a very impressive and decided stamp.

He brought a fine present of two bullocks,

several pots of pombe and milk, and some vegetable food. After the way in which he appears to have extorted a large hongo from Elton and Cotterill, I was astonished at the extremely small present which he was delighted to receive from me. It was the only place at which I ever gave less in value than I received.

To add to his other favours, he immediately constituted himself our guard, and strutted about proud of his shirt and his bells, laying violent hands on such of his subjects as were too inquisitive and rude.

It was altogether an extremely pleasing sight to see these simple-minded natives crowding round us good humouredly, to get the best possible look at us, with their frank and jolly chief thrashing them right and left, in the good-humoured fashion only to be seen in the harlequinade of a pantomime. Some had their faces painted in various designs, and wore feather head-dresses, while a few had got so far above the state of nature as to wear a bit of banana leaf. But even this concession was evidently more ornamental than otherwise, as instead of banana leaves many of the swells adorned themselves with bands of graceful creepers.

People who have never travelled, are apt to come to the conclusion that nudity can only be consistent with the lowest grade of human development; but this is really a great mistake. At least my experience among a varied collection of African tribes tells me so. Neither is nudity incompatible with modesty. Rather the contrary; for as a rule, I have found that among the negro races, the best dressed are usually the most immoral. Indeed, when we come to examine into the question, we usually find that it is more a matter of colour than anything else, and after that the forms of dress or undress depend on habit and custom. Nudity with a black skin never strikes me as strange or noteworthy; but with a white skin it is different. Indeed, I almost developed a morbid disinclination to look at myself, and once or twice I have even blushed to see my white skin. I became so afraid of being seen bathing even by my men, that it was only with the utmost secrecy that I ever attempted it.

The character of cleanliness which I first observed about the village, I found to be by no means outward show. All their domestic utensils, pots, &c., were kept scrupulously washed and free of dirt, and not an article was to be seen that any European might not have used without reluctance. As our acquaintance with them extended we observed fresh reasons for surprise. Every morning, the dead leaves which fell during the twenty-four hours were always carefully swept up over the whole village and burnt. Cattle were not allowed to come into the village, but were housed in most commodious and cleanly-kept quadrangular huts, which would have compared favourably with any similar building in Europe. Fowls were also kept outside.

Returning weary one day from a shooting excursion, I was also pleasantly surprised by the confidence with which the natives accepted our word. We had no cloth or beads, and we were dying for something to drink and eat. On our promising to pay them next day, abundance of milk was at once brought out, of which we got our fill. In almost every other tribe, we found it quite impossible to get the slightest article without immediate payment. So utterly ingrained is the habit of falsehood in themselves that they could not but distrust us.

Makula's people are not Wachungu as supposed by Elton and Cotterill, but Wakinga, who emigrated for political reasons from the high mountains north of the lake, and took possession of the fertile tract beside Nyassa. Formerly, from all accounts, Ukinga had been a country of considerable importance, till civil dissensions arose and broke up the tribe. Some then migrated, and some joined other tribes, till the whole district at last fell under the rule first of Merere, and then of Mamlè.

The hill on which they live has been described as a mountain of iron by Cotterill, but he would have been nearer the truth if he had said a mountain in the soil of which a few nodules of iron are found and worked. These people certainly make some exceedingly curious spears. When those now under Makula migrated, they carried the art with them, and continued making their own spears, though the raw material is still derived from Ukinga. Their spears (which are their only weapons) are barbed in the most frightful manner, so that any spear stuck in the flesh would make a horrid wound before it could be withdrawn. They seem most unaccountable weapons in the hands of such harmless people.

At the river Jumbaka, close to our camp, I had some good shooting at crocodiles and hippos, which swarmed in some of the deeper pools. At this season the river was so low that at most places it was only knee deep.

On the second day after our arrival a *fracas* arose which at first seemed to threaten serious trouble. I was in my tent dozing away an idle hour, when I was rudely awakened by a great uproar. Running out I observed a struggling mass of men, both of Makula's and my own, which was being augmented on all sides. Spears and sticks were being used in a very promiscuous and dangerous manner. My shouts were unheeded, and perceiving it was a case for prompt and decisive action, I seized the trunk of a banana which happened to be lying beside my tent. With this harmless, though formidable-looking weapon, I rushed at the crowd, and with all the energy of which I was capable I smashed right and left, laying low many a worthy warrior and porter, till I cut a pathway into the very centre of the battle, as was done in the goodly days of old, when gallant knights thought nothing of ploughing their way into the heart of an enemy's army.

This prompt measure at once scattered the crowd, and I had then time to inquire what was the meaning of the quarrel. It appeared that by accident one of my porters had stabbed a warrior in the eye with a spear, and then his comrades had attacked them. I at once sent for Makula, and had the matter explained, and a good present given to the injured man; and as only a few bruises had resulted from the fray this put matters right at once. If I had happened not to be present, the consequences would have been very serious.

Food of all kinds proved to be so abundant, and the natives so anxious to get bits of cloth, that I was able to get as much food for the men each day as was needed, for cloth valued in Zanzibar at less than two shillings. A strip of cotton from an inch to two inches in breadth, and just sufficient to tie as a band round the head, bought a day's rations for two men. Much to my indignation, however, as the natives were beginning to get clothed and in their right mind by our system of bartering, my own men were rapidly reverting to a state of nudity. Finding the climate warmer than in the mountains, and the food so extremely cheap, they could not resist the temptation to strip the clothing off their outer man that they might gorge the inner to repletion. Thus one bit went, and another, till many of them were little better than the surrounding natives. Then with long faces, and mournful, beseeching voices, they would come to me in the most pitiful cringing manner, saying, "Bwana mkubwa (master)! Look at us, our clothes are finished. They have become old, and fallen off in rags. Give us new clothes. Are we not your children? and must we go like the Washenzi?" and much to the same effect. But I was obdurate, and hardened my heart against them, telling them to get out of my presence at once. If they thought fit to fill their stomachs with their clothes they must needs go naked. Their foolishness they soon regretted, when they once more reached the top of the Plateau. But I had to be firm, for if I had yielded once, then the whole caravan would have been down upon me, and it would become an everlasting annoyance.

I would fain have stayed several days in this charming Arcadian village, enjoying the generous hospitalities of Makula, but I felt I must go at once if I did not want to be rendered unfit for travelling. In spite of all its charming appearance the village was unfortunately placed among marshes, which continuously emitted double-distilled malaria, which I felt I was imbibing to the detriment of my health every hour. A constant nausea and feverishness kept possession of me, and threatened the worst consequences. So I was perforce constrained to hasten my departure sooner than I otherwise would have done. The men were thoroughly disgusted at being obliged to leave such glorious fleshpots, and vented their ill-humour in loud complaints; but to start I was determined.

On the 29th of September we got under weigh for the second great stage in our march, Tanganyika. So little did the natives of Makula know about the geography of the surrounding country that the existence of Tanganyika was quite unsuspected by them. At least to all my questions I got only vacant stares.

Makula's major-domo, a tall ungainly character, clothed in a gorgeous dressing-gown and straw hat, constituted himself our guide through his country, and Makula himself accompanied us on the first day's march, bringing a bullock with him as a farewell present. I found I had not started a day too soon, as I felt so ill after starting that twice I had to rest for an hour at a time, of which my men took advantage to vent their ill-humour, marching a distance longer by half than I had ordered, and making me drag my poor limbs in torture after them. Our route lay along the banks of the Jumbaka, through a continuous grove of bananas with scattered houses. We crossed the river twice, and saw numerous evidences of its having frequently changed its channel.

Next day I felt somewhat better. We got clear of the banana groves, crossed a bit of fine grassy plain, penetrated a tract of wood, and camped at eleven o'clock at the miserable little village of Pokirambo.

In the bed of a small neighbouring stream we saw some fine sections of the volcanic rocks, which attain such a magnificent development in the mountains between Mtandala and Pupanganda, rising in magnificently picturesque masses to a height of from 7000 to 8000 feet, and consisting of porphyrites, and garnetiferous tuffs and agglomerates.

From Pokirambo we observed a curious conical hill of the most regular shape, rising abruptly from the plain. The cone was truncated, and it presented such marked characteristics that I at once recognized it as an extinct volcanic crater, probably of much later date than the volcanoes, which had erupted the great mass of igneous rocks that formed the mountains. It seemed clear that the plain on which the cone stood had been cut out of the subtending plateau first, and that then the country had shown signs of renewed volcanic activity, producing the small cone.

Next day my supposition as to the character of the cone was verified. It rose to a height of 250 feet above the plain. A pathway, which puzzled me at first, led up the side. This I took advantage of in my ascent. On reaching the top I was perfectly enchanted by the sight which met my eyes. There lay one of the most perfect volcanic craters that could possibly be conceived. The centre of the hill was cut into a circular bowl-shaped hollow, with smooth sides, and narrowing as it descended, till, near the level of the plain, the circle would be no more than sixty feet in diameter. The circle at the bottom formed a pretty pond, surrounded with bushes, and of unknown depth, inhabited by hippopotamuses, whose presence at once explained the existence of the pathway which I had been puzzled to account for. The rim of this beautiful crater was quite entire, and formed little more than an edge to the bowl, which at the top would be quite 200 feet in diameter. The perfect preservation of this crater is another reason for supposing that it must have originated in comparatively recent times. We can hardly imagine how it could stand uninjured by denudation through geological periods.

Several other cones could be descried around, but there was neither time nor opportunity to visit them. A few miles to the N.W. we passed a charming circular lake, without an outlet, and nearly a quarter of a mile in diameter. Its depth is unknown, but from its position and surroundings I am inclined to assign it to a volcanic origin.

If we now collate all the evidence we have got it will be found that there has existed a great line of volcanic activity stretching from the Cape Colony by the Zambesi, Nyassa, Ubena, Ugogo, Kilimanjaro, to Abyssinia. How far the origin of Lake Nyassa may be connected with this volcanic belt, I cannot take upon me to say, having as yet examined only the geological features of its northern termination; but that there is such a connexion will hardly be doubted.

After passing the volcanic cone we crossed the

Jumbaka, and passing through an extremely pleasant piece of forest, with ground gradually rising in altitude, we camped at the village of Mmasoka (probably the Mazotè's of Elton), in a small valley filled with bananas. As we gradually rose in altitude and got into purer air, the same effect was produced which I have already described as marking the change from the coast lowlands to the mountains. The malaria began to work out of me, and reduced me terribly. It can hardly be conceived how dreadful it was to pull myself over mile after mile of country, feeling that it must be done if I would live at all. At Pokirambo I was delirious during the night, and full of the strangest fancies. I supposed that all the men were making preparations again to desert me *en masse*, and that this time every man would go and not come back. In an agony of fear I dragged myself out of my tent at midnight, and called on Chuma not to forsake me, and to try his influence with the men. What had I done that they should thus leave me? &c. All this I learned from Chuma in the morning, when I got over the delirium, and walked nearly ten miles.

Each day I usually contrived to pull myself together for the march; but I was just like a machine, wound up to go a certain time, and then collapse utterly.

I began to think that after all the natives were not so far wrong in their belief that disease was caused by devils getting into one. I certainly often felt as if I was possessed with a number of these unwelcome gentlemen, so excessively irritable did I become under my complicated troubles. I felt like an untamable animal newly caged, and in my impotent rage would almost dash my head against any obstruction. The weaker I grew, and the more unable to wreak my vengeance physically on any of my men who thwarted me, I developed a certain strength of language which would have astonished those who know me. It was, however, a convenient safety-valve for my feelings, and after hurling a string of epithets at the devoted head of some one of my men, I felt that I had done my duty, and was proportionately relieved in consequence. I make this confession in justice to the Zanzibar porters. We hear frequently about their troublesome conduct, desertions, obstinacy, &c. But we are never told how much they have to bear from their masters, when they have lost all moral control of themselves by incessant troubles, and throw, on the most frivolous pretexts, sticks, books, boots, or the strongest language of Billingsgate, at them. My cook, Litali, who had been brought up at the Universities' Mission and taught where bad boys go to, felt constrained to admonish me on one occasion, when I had felt it necessary to express myself forcibly about the way in which he was neglecting his duties much to the detriment of my constitution.

From Mmasoka we obtained a fine view of the so-called Konde mountain as figured in Elton's book.

Our route had up to this point run parallel with the volcanic mountains, which trend about north-west and south-east for nearly thirty miles, and then sweep round and join the range of mountains running north and south between Uchungu and Makula's country, which thus occupies a triangular niche cut out of the surrounding Plateau—for when I speak of mountains, it would perhaps be more appropriate to speak of the escarpment or edge of the Plateau.

From Mmasoka we turned more to the west, and after crossing two small streams flowing to the Jumbaka, reached the basin of another and larger river called the Lukuvira. This river runs along the base of the Uchungu mountains, which it drains, and it falls into Nyassa south of Makula's country. This part is inhabited by a small sub-tribe called Wakukwe, and credited by Elton with the character of dangerous robbers. If robbers they can be called, they are certainly the most cowardly that can well be conceived.

Although we took every precaution to intimate our approach and our friendly mission, we found every village deserted on our route. Every here and there could be seen villagers flying for their lives in all directions, carrying what they could with them, and driving helter-skelter their cattle before them. Away on the high mountain peaks groups of the villagers could be faintly discerned, watching our movements, and doubtless expecting to see a grand conflagration of their villages. In almost every case, however, we succeeded in establishing confidence, and bringing the villagers back. This was most frequently done by capturing a native, and showing him that not a house was touched or an article removed, thereafter giving him a present and sending him as an ambassador to his brethren.

On the 5th of October, after crossing a much-broken country, we forded the Lukuvira, and ascending one of its tributaries by a most charming valley which led into the mountains, we camped at Mwisika, on the top of a ridge, amidst scenery of the most picturesque description. Away down the valley, as through a telescope, we could see the great plain of Makula, with Nyassa beyond. On our right rose smooth grassy mountains, terminating in peaks which were occupied by Wakukwe cowards in picturesque groups. Behind, the mountains swept round us like the half of a great ruined amphitheatre, with frowning precipices and dark green forest patches. On our left were more hills, dotted over with the pretty villages of the natives embosomed among bananas, with dashing silvery cascades, terraced patches of cultivation, and grassy slopes. At our feet lay a deep narrow gorge, through which a noisy stream wound its way, sending up a plaintive sough like the great sighing forest when a gentle breeze wafts over it. Such was the delightful position of our camp on the 6th of October.

In the evening we were surprised by the sight of

a line of men with shields and spears descending the mountains, appearing and disappearing according to the nature of the ground. Gradually they approached, and their steady march excited our curiosity to know who they were, as they certainly were not Wakukwe. At last they came near the village, and without the slightest hesitation they marched in as if they were lords of the country. They carried shields of hide, and spears of the Wahehe type. They all wore loin cloths of native-made cotton—the first we had seen—beside feather and hair head-dresses. Their boldness and absence of fear, however, took us most by surprise. On inquiry, we learned that they were Wanyika, a tribe through which we would have to pass. They had heard that there was war on the road, and they had come to learn the truth of this rumour. They were exceedingly haughty, and would only answer such questions as they pleased, and they refused to be our guides. This was rather ominous, considering that we had to pass through their country. If they all proved to be of a similar nature, we might anticipate trouble of a more serious nature than anything we had yet met among more cowardly tribes.

Next day we had a very stiff climb of nearly 4000 feet up the face of the escarpment, through a dense piece of forest, which, in its characteristics, reminded us of the forest slopes of Usambara, only there were more flowers, and they sent out a strong honey odour, which was most sickening. On reaching the top the view was not encouraging. It had a most irregular aspect, with ridge succeeding ridge, in great part smooth and grassy, though occasionally a patch of trees varied the otherwise monotonously grass-green colour. Over this we tramped for two hours.

About three in the afternoon I found myself considerably ahead of my men, with only my gun-bearer. We reached the village of Pamalila. In front of it we found several Wanyika, who looked somewhat startled by our strange appearance. Unlike the Wakukwe, however, they simply stared their astonishment, and then began to walk leisurely into their village. I courteously saluted them, but they did not reply. Not heeding this I followed them, with the intention of getting some shelter from the hot sun till my men arrived. As I was about to enter through the gate, smash went the huge log in my face, almost striking me, and within an ace of crushing my arms between it and the gate-posts. As it was I left a bit of my sleeve in the place. A derisive laugh from inside greeted my discomfiture, which made me boil with rage, and vow to storm and burn the village. However, I thought better of it, and retired in disgust to the shade of the neighbouring forest, where we camped.

The village which thus bearded us could not have mustered fifteen fighting-men. Still we considered they had a perfect right to keep us out if they pleased. So they escaped feeling the power of our guns. On the contrary, we soon proved to them that we were friends and could be relied upon; and they finally opened their gates, and bartered food for cloth.

The tribal differences between Makula's people and the Wanyika were of the most marked sort, not only as regards their mental and physical features, but also their habits and customs. Of their mental difference we have already given a few examples. Physically the Wanyika were not so good-looking, having long lank bodies, with exceedingly narrow craniums. Hair, on the breast, face, and head, was exceptionally abundant,—in this respect they almost rivalled Europeans. They allow their hair to grow uncut, and usually it appears as enormous greasy masses. They wear a loin cloth of rough cotton, made by themselves, and resembling the very coarsest sacking. They are extremely filthy about their villages, presenting in this respect a very disagreeable contrast to Makula's people. Owing to the almost constant state of warfare in which they live, the Wanyika are compelled to live in stockaded villages. The huts are huddled as closely together as possible, leaving barely room to creep about among them. The area to be defended is thus lessened. At night their cattle are brought within the stockade, filling up all the odd spaces; and as the filth is never removed, the frightful condition of the interior of a Wanyika village may be conceived. Personally they have no delicacy or cleanliness; hence the ground outside their gates is one great dunghill. This filthiness extends to everything else—the insides of their huts and their domestic utensils. The huts are very poor specimens of the kind. They are generally falling to ruins, and are badly built. But for their utter want of union among themselves the Wanyika would certainly form an exceedingly important tribe. For boldness and cool impudence they have not their match in East Central Africa. Though they had hardly even heard of a white man before, yet they scrupled not to defy him and his caravan, who outnumbered them as five to one.

The day after our arrival, we had even a more amusing instance of their impudence than we had yet witnessed. We were marching along over hill and dale, not suspecting any trouble, when, behold, six warriors with their chief commanded us to halt. He wanted to know what business we had in his territories without asking his leave. We must go back at once as he was determined we should not pass his road. This was sufficiently alarming, so I commanded my 150 men to halt till we had discussed the question. After a deal of talk and the offer of a small present, this African lion was soothed without bloodshed, and we were allowed to pass peaceably through his village of six huts, his warriors being drawn up in line, past whom we marched with all the honours of war. Camping beside a small village called Mwizombwe, we were not allowed to enter, and as there was no cleared space, we had to find accommodation as best we might, among dense brushwood. The spectacle presented by our camp was most amusing, as the porters ensconced themselves in the heart of the bushes, cutting the centre out for that purpose, and appearing very like comfortable monkeys.

Hardly any food was to be got, but fortunately honey was to be had in extraordinary abundance, and it formed an agreeable and a nutritious substitute. The men enjoyed the young bees as much as the pure honey. Nyika appears to be a great bee resort; the very air seems filled with one continuous buzz, and the odour of honey was very strong.

Next day we ascended the Munboya Mountains, attaining the altitude of 8200 feet, the highest point crossed in our journey. At the village beside which we camped, we found the natives prepared for defence, and all the pathways leading to the village blocked with thorns. In the evening I tried to reach it, but after tearing my clothes to rags I had to give up the attempt, without getting even a sight of Mtanda.

At this point a stream rises, and flows first north and then west to Lake Hikwa, or Leopold, as I have presumed to call it. To the south, however, streams also flow to Lake Nyassa.

Few incidents marked our progress through Nyika, beyond the ordinary troubles which always dog the explorer's footsteps in East Africa; troubles irritating enough to the traveller himself, but possessing no particular interest to the reader of his story. I may, however, relate an incident illustrating the shifts to which a caravan leader has to resort, to keep his men in order. Few people will be inclined to think that castor oil may, under certain circumstances, be exceedingly useful in oiling the wheels of caravan life. Let me show how I used this simple and efficacious family medicine to break up organized conspiracy.

When the idea of having a day's rest took possession of the men, they were in the habit of manufacturing as many sick men as possible, and then insisting that it was impossible to get on without a day's halt, to give the invalids an opportunity of recovering. One morning I observed this dodge in process of execution. Loud grumbles of "Si wezi, si wezi sana" (I am very sick), were heard on all sides, and a large number congregated in the rear, declaring they were unable to carry loads. After much trouble we got the bales arranged. I took no notice of these preparations, and in my usual manner marched away in front. I was not surprised to find myself an hour before my men in camp, or to observe a miserable string of porters marching in the most woe-begone fashion, holding themselves as if they had the gripes, and groaning in the most heartrending manner and looking inexpressible agonies. I at once saw that there would be no march next day if they were not instantly cured. But I was determined to get on, and I resolved to put an end to this oft-recurring nuisance. Calling up the sick men, I asked in the most sympathetic tones at my command what their ailments were? They all tenderly rubbed their stomachs with a lugubrious chorus, "Tumbo bwana, Tumbo mbaya sana" (My stomach, master, my stomach is very bad). I rejoiced to hear that that important, though vulnerable, part of the constitution was out of order, as I had a very simple remedy. Smiling benignantly upon them, I told them to be of good cheer, as I would soon put them all right. Going into my tent, I brought out two large handsome bottles of castor oil. Now mark the effect that the mere sight of that simple maternal medicine had upon these broken-down creatures! Groans were hushed. Their hands dropped from the affected part, and every one tried to look all right, though a glance of alarm passed from patient to patient—for be it understood, there is nothing a native detests more than European medicines. Strange and marvellous to relate, they were all inclined to retire, cured by the very sight of the bottles. This, however, was not going to suit my purpose. So with parental sternness, I ordered them to sit down and open their mouths. Not without difficulty, I administered a large dose to each man, letting him know that if it did not prove an immediate specific, the dose would be repeated next morning. There was no dancing or singing in the camp that night, though many of the men were astir. Next morning every man was convalescent, and on my doubting the fact in some cases, they at once rushed off and showed me how vigorously they could pick up their loads. Thus not having the usual excuse for a halt, we marched out of camp most merrily.

The same dodge was never repeated; a fortunate thing for me, as the castor oil was finished, though luckily the men did not know it. It has always been a subject of regret to me that castor oil is not a patent medicine, as then my testimony to its immense value to travellers would probably have brought me in a nice sum of money.

I, however, did not always get over my troubles with the men so easily. I was so unfortunate as to quarrel with Chuma and the other headmen. I suppose we were equally to blame, for in my irritable frame of mind I was not always able to command my language, and very apt to rush hastily to conclusions. For two weeks at this time Chuma and I were at daggers drawn, as he felt himself very much aggrieved, and showed me very conclusively his power of annoying me. This made me miserable enough, though frequently the incidents that it gave rise to were sufficiently ludicrous to afford me great amusement. At last we got matters smoothed over, and we took care not to quarrel again.

At the village of Kwachuma we had an instance of the way a petty chief can obstruct a caravan. We were anxious to get along, and consequently informed his lordship that we proposed going on next day. He returned for answer that "he was a chief, and was not to be forced; he must have time to collect food as a present, and get guides for us. Had we not stayed a day at the last village, and was he to be passed in this unceremonious fashion?" He was firm, and would not yield one iota of his privileges, and to stay we were compelled.

It may seem absurd to many that an insignificant chief of a few warriors should compel a well-armed caravan to bide his pleasure; but in Africa a petty chief has a wonderful influence. If we had resisted his demand he would at once have spread the report that we were forcing our way through the country, and that we were enemies. The consequence would have been the rousing and alarming of the whole country in front; villages would have been deserted, food unattainable, guides not to be got; and, finally, open war would have resulted. This might have been exciting, and would have made our march sensational enough. But we had not come to Central Africa to reap a crop of exciting adventures; we prided ourselves on being pioneers of civilization, and we came armed with peace and friendship. In dealing with savages therefore we saw that it was necessary to be gentle, and humour them in their ideas and customs. We claimed no right to force our way.

As we travelled westward a change in the appearance and character of the Wanyika was noticeable. They became less dark in colour; figures and features improved, and hair on the face or body seemed less abundant. Their villages became larger and cleaner, while copper and brass ornaments, together with beads, were more frequently seen on the natives; cattle disappeared, and were replaced by flocks of goats and sheep. Food was more plentiful, and therefore cheaper.

The country, in its general aspect of grassy ridges and scrubby tracts, remained the same throughout Nyika. It was about the close of the wet season when we traversed the country, and not a green blade of grass was to be seen. Most of the trees were shrivelled up, although some of the bushes were putting on new shoots of dark green, relieving to some extent the dreary yellowness of the rest of the vegetation. The same absence of animal life, which we noticed as marking the plateau of Uhehe and Ubena, characterized Nyika. Neither bird, beast, creeping thing, nor insect was to be seen, except on very rare occasions. We felt it to be quite a memorable day when we shot an antelope at Kwachuma; it was actually the first specimen seen since leaving the valley of the Rufiji, to the east of Behobeho.

On the 16th of October we reached the western

boundary of Nyika, at the village of Mtinga. A considerable stream here flows northward to Lake Hikwa, or Leopold, which is only some twenty miles off. I endeavoured to get guides to it, but failed, as the country around it was in the hands of Merere, with whom they were at war. This did not trouble me much, as I proposed on my return march to explore it.

On the 18th we crossed the Chingambo Mountains, and entered a new country called Inyamwanga. On camping in the afternoon I selected a very fine shady grove, where I ordered my tent to be pitched. While this was being done we were startled by a tremendous outcry. The villagers came rushing in from all sides in the greatest excitement, ordering us to desist. It then appeared that the place was sacred to their late chief, who was buried in a hut close by. We did not want to shock their feelings, but entered into negotiations with them, as I was unwilling to remove out of the cool shade. I informed them I was willing to take the risk of being killed by his Majesty's ghost if he should walk in the night, but hinted that perhaps a timely present of cloth would arrange the matter to the satisfaction of the late chief. This compromise was accepted, and the then chief took the cloth to the grave. Before entering the hut he laid aside his knife and spear, as did my men when they came near. He then commenced a long speech, informing the ghost that a white man had arrived and pitched his camp beside him. He had, however, piously vowed a cloth to the repose and comfort of the dead, and hoped that no evil would befall him for his temerity. Whereupon the living chief neatly stowed away the present about his own person and retired, after some mumbling, which I suppose would be the ghost's reply. I do not know whether the tsetse fly has anything to do with released spirits, but I must confess they plagued me that night in a maddening manner.

Next day we arrived at Mswilo, the town of Mlilo, the chief of Inyamwanga. I found it to be of unusual size, with the aristocratic portion divided by a palisade from that of the lower classes. I took up my quarters in the centre of the village, and tried with much comfort to myself the cool virtues of a native hut. We arrived early in the forenoon, and as the chief was not to be seen that day I occupied the time taking stock. I found that I had still half of my cloth and wire, and about all the beads. This encouraging position resulted entirely from the determination with which I stuck to the old system of buying food.

Up till within late years the only mode of rationing the men was by the master of the caravan buying each day's food himself, and then measuring it out in a dish of the required capacity, called the kababa. Of late, however, among European caravans this system has been changed. Instead of buying and distributing the food, the caravan-leader now measures out to each man so much cloth to purchase his rations for a certain number of days. At first it commenced with two yards for fourteen days to each man. Then it gradually increased in amount till each man now with European caravans receives two yards for four days' food. Hence 100 men would require 200 yards every four days to buy food.

Now from the very first I determined to stick to the old system, as infinitely more cheap. The men, who knew the advantages of the new system to themselves, of course frequently demanded to have it applied to them, but I was firm. It was only too clear that if I gave in it would be utterly impossible to go beyond Lake Nyassa. By the time we reached that point our stores would be quite exhausted. I found that by buying the food myself I could feed 100 men one day for eight yards of cotton, and consequently four days for thirty-two yards. The new system, as I have shown, required 200 yards to do the same, being more than six times as much. By thus keeping to the kababa I was enabled to travel over three times as much country as I otherwise could have done.

This vast increase in the cost of feeding a caravan is rapidly becoming a great obstruction to the exploration and opening up of Africa. A porter now almost eats up his load in four or five months. Consequently caravans are becoming necessarily larger and larger each year. Europeans have themselves to blame for this, though I suppose the native will have to bear the odium. Europeans introduced the practice to save themselves a little trouble, and it has more than doubled the expense of travelling in Central Africa. I suppose our big Missionary Societies, and the International Association, are not particular about the matter of a few thousand pounds, and will comfort themselves by the thought that the money is at least spent among the benighted natives of Africa, who may by the abundance of cotton be induced to clothe themselves decently, thus opening a door for putting them in a proper frame of mind.

In the evening of my arrival at Mswilo I was much troubled by a blind native musician, who sat down in front of my door surrounded by a sympathizing crowd of natives and porters. He shrieked away in the most extraordinary cadence, accompanying himself with two calabashes containing small stones, which he rattled with deafening effect. He seemed to interest his hearers very much, and as he was blind I tolerated him for some time, although I did not understand what he was singing about. At last, getting exasperated, I ordered Chuma to make him move on. "Oh!" said that gentleman, "you must give him something. He has been very badly used by Mlilo, who put out his eyes because he was found at late hours in one of his wives' huts. That is what he is singing about. He now gains his livelihood by recounting his deeds of love and conquest."

"What!" said I, "do you expect me to encourage such proceedings? Clear him out at once!" Chuma looked surprised at my want of sympathy, and did not move; so I settled the matter by throwing my camp stool at the head of the aggravating old sinner, which stopped his vile singing, and left me in peace.

The following day we were graciously informed that Mlilo was ready to receive us. Taking my head men, who always formed my guard of honour on these occasions, I proceeded towards the palace (if his hut might be so denominated). With much difficulty we threaded our way through the west end of Mswilo, in continual danger of getting a projecting rafter in the eye or through the skull. The chief's house and his wives' huts we found to be cut off from the rest of the village by a strong palisade. Outside were situated his majesty's breweries, where pombe in huge pots could be seen in all the stages of its preparation. Passing into the enclosure we observed the chief's hut in the centre. It proved to be of the usual type, but of unwonted size. The huge conical roof projected all round over the walls, forming a capacious verandah, where shade and the cool breeze could be enjoyed. Into this we were ushered by the master of ceremonies. I took my seat on the camp stool I used on these occasions.

While Chuma recounted who we were, where we were going and what our objects were, I had time to take in the scene. In front of me sat Mlilo on a high raised seat, swathed in gorgeous cloths, his legs being covered with brass and copper leglets and grass cords. Huge strings of beads adorned his neck, and his arms were profusely adorned with a variety of ornaments. He is old, tall, and very stout and bloated-looking, with bleared eyes, as if perpetually intoxicated. Altogether he looks the very type of a savage potentate. On each side of him stood a pot of pombe, in a beautifully made basket with a cover. In front of him sat a little boy, who held the tube through which he ever and anon recreated exhausted nature by sucking up the pombe, on which he almost entirely lives. On my right sat the courtiers, looking properly abject and humble before such greatness. On my left were Mlilo's wives, warm brown buxom creatures, with smooth shining skins, and eyes that spoke of intense amusement and curiosity. Here and there the more bashful could be seen peeping, like frightened rabbits, from behind doors or round corners. Entering at once into the fun of the thing, I forgot I was in the great chief's presence, and winked at the prettiest. The giggling and nudging of each other at this intensely humorous act attracted the attention of every one, and looking round I was taken aback at seeing the stern face of Mlilo, the frightened look of his courtiers, and the amusement of Chuma and the other head men, which made me blush when I thought of the camp stool and the blind man.

Recovering ourselves, Chuma finished his speech; and Mlilo replied in a very dignified manner, expressing his pleasure at seeing the white man, and the many strange things he had to show. He was pleased with our present, and guides would be ready for us whenever we were ready to march. After the winking scene the women had all disappeared in a fright, and during my stay in Mswilo they were closely confined—a highly unnecessary proceeding, I thought, and reflecting injuriously on my character.

About Inyamwanga and its people we need say nothing, as they present no peculiarities worth noticing. All the streams in this part trended south, forming the headwaters of the river Chambeze, which falls into Lake Moero, and thus forms part of the headwaters of the Congo.

Our march was unrelieved by any incident worth relating, till we reached its borders, at Misogwere. What might have been a serious calamity befell us here during the night. It was very dark and gusty. The fires burned apparently with increased brilliancy, contrasting with the outer gloom, and casting deep shadows. To protect themselves from the cold wind and threatening rain, all the men had constructed grass huts in a circle round the loads and my tent. About an hour after I had fallen asleep I was suddenly awakened by a variety of excited yells, and a great glare of light. The camp was on fire. Seizing my coat, I rushed out to see a most magnificent conflagration, with an excited crowd of men in their white kanzus, making the confusion ten times worse, and apparently adding fuel to the flames. I was quite astounded at first, but soon recovered myself. In the centre of this fire, and in the midst of 150 men, there were twelve kegs of gunpowder and all our goods. We were certainly in imminent danger. Chuma, forgetting about the powder, was rushing about trying to direct the men; but I yelled to him to look after the baruti (gunpowder). Rushing to the loads, I seized the first lot of kegs which I could lay hands on, and observing a small opening in the circle of fire, I made a rush through it with my load, and put it in a place of safety. Chuma and Makatuba followed with the others. Then we rushed back to save what else we could. The scene was of the most exciting character. The men, awakened out of sleep, had not had time to think of their guns and personal effects, and now they were rushing madly into the flames to save what they could, despite the exploding of the loaded guns and the danger from the powder-horns. The headmen had all their effects round the heap of bales where they usually slept, and round these they kept guard, dashing out the fire as it advanced.

Meanwhile my tent, with its precious contents,

had been forgotten. Returning to it, I was horrified to see the flames rapidly creeping towards it. Another moment and it would have been on fire; but my coat was ready, and in spite of burns and smoke I thrashed away with the energy of two men and soon got the mastery. Putting some men to guard it from further danger, I had time to look about me.

The worst appeared to be past, and the bales at least were all safe. A large number of the men, however, had lost everything belonging to them. In some cases the clothes were burnt off their backs. Some of the escapes were truly marvellous. The guns are always kept loaded at night. When the panic got up, these were deserted. Then in the attempt to secure them again, several exploded with the heat when got hold of, but fortunately without damage. One or two of the powder-horns also exploded, knocking a number of the men down. One of the most alarming episodes was the sight of a stupid fool issuing from the flames with one of these horns in a dangerous condition. Fortunately it was instantly torn from his grasp and whirled through the air, exploding with violence the next moment.

At last the fire went out, and in another hour we were enveloped in total darkness. Next morning the camp presented a rather woe-begone appearance. Fortunately, however, the loss was trifling, consisting only of a few clothes belonging to the men, which were easily replaced. But while it lasted, a more exciting scene could hardly be imagined. For a few minutes we were in imminent danger of being blown to pieces by our powder, and how we escaped without any serious casualties I cannot yet comprehend.

On leaving Inyamwanga, our route to the lake lay through the district of Mambwe. We found little to interest us here, the natives and country being similar in all respects to those we have already become acquainted with.

Signs of the approaching wet season now began to appear, in fleeting clouds, occasional showers, and almost daily thunder. This is one of the two busy seasons in the African year. The ground was being prepared to take advantage of the coming rains. Huts were being re-thatched and repaired, and a general air of industry pervaded the natives, who could hardly spare sufficient time to come and stare at us.

The people of Mambwe employed a wonderful variety of materials for the purposes of clothing. There were native cotton cloth, imported European cotton, goat and antelope skins, bark cloth, and, most curious of all, the stomach of the elephant.

Elephants were formerly very plentiful in Mambwe, but they appeared to be quite exterminated when we passed. The spears they used in the chase of the elephant were of a monstrous size, forming almost a man's load; it seemed as if they would require giants to wield them. The shaft is short and thick. They have always attached to them pieces of snake-skin, as a good elephant charm, by which they ensure the death of the animal.

We visited Mambwe, the chief, at his village of Muluchuchu. We found him to be a greedy, besotted young man, of the most cowardly character. He dared not stand, and see me shoot with my gun.

The country to the north-west of Muluchuchu is formed of a most beautiful plain, from which several of the small tributaries of the Chambezi take their rise. The ground was quite level, and the grass, for a wonder, quite short and sward-like. Numerous huge ant-mounds dotted the plain. These mounds seem to favour vegetation, as they were usually occupied by dense brushwood, with a tall tree at their tops. Here and there belts of forest diversified the landscape. Lily-covered ponds were not wanting, and the whole was enclosed by low wooded hills, with the Uembe mountains in the background.

This plain I have no doubt becomes a stagnant lake during the wet season, which accounts for the presence in the soil of numerous nodules of iron; this is worked by the natives, who smelt it in well-made furnaces. Most of the rivers and springs are charged with iron in solution, which, under proper conditions, is precipitated as bog iron. It generally takes the form of nodules, and it is from this source that the natives over the whole of East Central Africa obtain the iron which they work, and which has given rise to the story of the fabulous mineral riches of the country.

At the northern extremity of Mambwe, and at the base of the Uembe mountain, the natives informed us that two large streams took their rise from a great spring, and flowed in different directions. One trends south, and forms the main artery of the Chambezi; the other flows west, and falls into Lake Leopold. This latter stream, called the Saisa, we crossed about half a mile from its source. We then found it to be armpit deep and twelve feet broad, so that the spring from which it rises must be of unusual size. It is certainly a very interesting fact that the streams from one spring should take such different courses.

On leaving Mambwe we entered the country of Ulungu, distinguished from the former by its hilly and more varied character. On the 31st of October, after a very long march, we reached the town of Sombe, one of the chiefs of Ulungu. As we entered we felt we had reached classic ground, for we had now got on the footsteps of Livingstone. We here found two Mlima Arabs, on their way to the coast with ivory and slaves. The latter they kept carefully out of sight, marching them off in the dead of night, and following with the ivory in the morning.

It soon became only too evident that we had arrived on a trade route. The people were much better dressed, and even the children sported a rag. Beads and other ornaments were unusually common. The natives were exceedingly rude, and morality was at the very lowest ebb. The women signalized themselves most in this respect. The manner in which they crowded round me was most exasperating, more especially as I was suffering from toothache and other troubles. They watched my every movement, and made fun of me to their hearts' content, until I writhed in an agony of irritation. It was of no use to put my men to keep them away, it only became the merrier for them. At last, losing all control of myself, I threw boots, books, camp stool, or whatever came handy, at their heads.

I had taken possession of a half-finished house, in preference to my wretched tent, as the night looked promising. But about midnight I had reason to regret it, as the rain came down in torrents, and poured through the roof in spouts. In this condition I sat through the rest of the night, watching the blinding flashes of lightning, and listening to the thunder crashes. In the morning I felt nearly half dead, and I required some very hot tea to put me in order again.

Sombe turned out to be an exceedingly mean character. We gave him a good present, but he declared himself to be so poor that he could not give us anything in return. As if wanting to show, however, that he was doing his best, he marched about the village in the most ostentatious manner, with a piece of the blue cotton I had presented, and pretended he was trying hard to buy us a goat or a sheep. It would certainly have been rather an unusual thing to buy a return present with the present given.

In the evening the great event of killing a leopard was marked by immense rejoicing—speechifying, singing, drumming, and dancing.

Sombe has a few cows, the only ones to be found on the south or west of Tanganyika. We could not prevail on him to sell us one, and they were not then giving any milk.

After a day's rest we proceeded on our way. We crossed a hilly wooded country, and forded two streams swollen by the rain of the previous nights. We might have easily reached Tanganyika, but as there was no village there, we thought it better to restrain our impatience, and camp for the night at Mswilo, a village placed on a most remarkably steep slope. The people, we noticed, had all their front teeth filed to points, and tattooing was common. The women made no scruple of deserting the village *en masse* with my porters, whose actions in this respect I had no power to prevent.

In the afternoon I tried to get a glimpse of the lake, but failed, owing to the haze. The men came in a body asking to be allowed to fire off their guns when they arrived at the lake, and also wanting a present. I at once assented to the shooting, and promised the present, much to the satisfaction of every one.

On the morning of the 3rd of November, long before the sun was up, a general hum of excitement pervaded the camp. The porters required no shouting, beating of drums, and braying of horns to waken them up on that day. They all felt it was a day to be marked with white stones in their history. For when had ever Waswahili or Wangwana reached the Great Lake before in such a wonderful way, and so successfully? At last the day broke, Chuma gave the signal to the first Kiringosi, who tooted out a few notes on his pan-pipe, which were then taken up by the plaintive barghumi or horn. Next the drum, vigorously thumped, rolled forth its sonorous sounds, followed by the screeching of the zomiri. Finally, from the entire caravan a shout of "Tanganyika! Tanganyika!" burst forth, and every man sprang to his feet. In a twinkling tents were down, and loads shouldered; and before the sun bid us good morning we were *en route* for the shores of the lake.

Half an hour's walk through the thin forest brought us to the edge of the great lake crevasse of Tanganyika, and the scene that burst upon us seemed almost like fairyland compared with the monotonous features of the country we had just crossed. We had arrived at the most southerly point of Tanganyika, where as fine a prospect is presented as at any other place on the lake. It here forms a narrow acute angle running into the subtending plateau. On the left a point of land extends into the water like a huge quay, rising to a height of 300 feet in sheer rocky precipices, and topped by a dark green covering of trees, contrasting with the grey and the red of the sandstone cliffs. Over this could be discerned the placid Bay of Pambete, with its surrounding of picturesque mountains, from 2000 to 3000 feet above the lake, presenting a charming variety in the level bordering strip at the base, the lower rocky talus, and upper precipices, over which two beautiful cascades could be seen falling like silvery films. Over the point, three emerald islands showed themselves. On the right, or eastern side, the shore takes a fine sweep round to the north, and rises in steps and steep inclined planes covered with a dense vegetation, and culminating in the grand Lambalamfipa Mountains, 8000 feet above the level of the sea. At our side a delightfully clear stream dashed, in joyous babbling by pool and cascade, till it tumbled laughingly into the lake. From our feet extended the Tanganyika in expansive beauty, with its broken shore-lines and threatening walls of rock, its capes and islands, here sweeping round in a fine bay, there forming a miniature fjord. As we stood and looked, enchanted by the scene, the morning sun getting above the mountains and clear of the enveloping clouds struck the water, and from its rippling bosom the rays were reflected, forming a veritable field of gold, in which the green capes and islands lay like emeralds. The air was cool, and pervaded by a strong odour of flowers, which grew in profusion about us. The chirp of a few birds was heard, and waterfowl were to be seen in abundance on the lake. We felt as if we had passed from a purgatory to a paradise, so complete and glorious was the change.

On first seeing the lake, the men had taken full advantage of their leave to fire their guns. It was a proud moment for all of us, and we gave free vent to our delight in the manner of boisterous schoolboys. Down went the loads, and with shout and song a ring was formed of energetic dancers, who literally ploughed up the ground as if they were shod with iron, while every now and then the woods re-echoed with the roar of a gun.

After they had thus let off the effervescence of their feelings, we descended by a rocky pathway to the edge of the lake. Here we halted. The roll, which had not been called for a month, was now brought out, and in that solitary spot, with the lake rippling at our feet, each name is sung out, and a sonorous "Eh wallah" (Here, sir) is returned until the list is completed. Out of the 150 men who had left the Indian Ocean, there was not one absent. Neither desertion nor death had deprived us of a single porter—an occurrence unique in the history of African travelling.

I of course made a speech on the occasion, pointing out the great and unusual feat they had performed, and how much reason they had to be proud of themselves. This roused great enthusiasm, and they unanimously declared they were ready to follow me wherever I went. Each man then defiled past me, and shook hands with many pleasant and encouraging words, which made me feel that, whatever might be the colour of our skin, there existed no barrier between us, nor any difference but that of degree between our respective feelings and sentiments.

We had still some distance to go before reaching a camping-ground, but we could not resume our march till we had thoroughly enjoyed a good splashing in the lake and an hour's rest. Finding the sun getting well up in the heavens, we were at last constrained to move on. With much difficulty we climbed the cliffs on our left, and then crossing the rugged top of the cape we reached the large stream of Lonzua, which even at this late period of the year presented a swift and formidable current. On attempting to wade it I was swept off my feet, and carried some distance before I was captured by my men, who on seeing the catastrophe had dashed in after me with a will. Though somewhat chopfallen, I got off with nothing worse than the dipping. I could not help wondering at the curious power the natives have of keeping their feet in a current which I found quite impossible to resist.

We found a village on the northern bank, but the villagers declared they had no food, and directed us to Kasangalowa, further on. After a weary drag under a vertical sun we reached this village, and found it to be only a few huts. The rest of the afternoon I employed in distributing a present of cloth among the men, as I had promised; an action, which, I am sorry to say, I had reason to regret for a week after, as the men bought up every fowl, egg, and fish in the surrounding country, till I could not get one of these articles myself.

In the evening I had some exciting shooting. As darkness began to set in a considerable number of hippos came out of the water to feed on shore. Taking advantage of the darkness and the abundance of good cover, I got between them and the water, and gradually approached close to them unseen. This was sufficiently exciting in the dusk, and knowing that the huge brutes would make straight for the water by the place I was standing, I fired with a somewhat shaky hand. Whether I struck the hippo or the ground I do not know, but the moment the smoke cleared I became only too well aware that the hippo was bearing right down in my direction, whereupon I fumbled in my pocket for cartridges in vain, and then ignominiously made a dive into the centre of a clump of sedges, tripping over a creeper which brought me down half stunned on my face. I was only conscious of the crashing grass and the shaking ground, as the hippo went past a few feet from me, and I had the pleasure of seeing it splash into the lake as I rose. I felt inclined to throw my gun at my man to let off some of my chagrin.

Next day we moved on to Pambete. The caravan went by the ordinary pathway, but I had the "Agnes" launched on the lake, and in the cool morning rowed round with my two boatmen. I had a few shots at some crocodiles, which are here very abundant, but was perfectly startled by the thundering reverberation of our guns, which seemed to sound like the simultaneous firing of a whole field of artillery.

Two hours' rowing round a small headland brought us to the village of Pambete, memorable as the spot where Livingstone, in 1867, reached the lake. We here resolved to stay a few days to consider carefully our future plans of action, as we had now arrived at a very important point in our journey.

Before proceeding any further with our narrative, it will be conducive to a correct and clear notion of the country between the two lakes if we take a rapid retrospective glance, and gather up the various fragments scattered at large through this chapter.

It will be remembered, that starting from Lake Nyassa we passed in a general W.N.W. direction through the low-lying country of Makula, with its magnificent grazing-plains, its marshes, miles upon miles of wonderful banana groves, its fields of sugar-cane, cassava, sweet potatoes, millet, yams, beans, Indian corn, in the richest profusion. To the west rose the steep escarpment of the Plateau, met by the magnificent volcanic mountains of Ukinga and Konde in the north.

As we proceeded westward we rose in altitude, leaving behind us the malarious marshes, but also the bananas and other vegetable productions. We found the country more bare of trees, but affording splendid grazing-ground, which was taken advantage of by the sub-tribe called Wakukwe. At this part numerous streams from the mountains to the north cut the country into a succession of narrow gorges, and flow to the river Lukuvira, a large stream which goes to the lake. Passing up one of the tributaries, we get among the mountains again, and finally reach the general level of the great Plateau.

Here again we find a poorly wooded country, very much cut up by numerous streams, but affording good grazing-ground. At the Munboya Mountains we attain the height of nearly 8200 feet. From this height to the village of Mtinga, on the borders of Nyika, the land descends in height till its altitude is but 3000 feet. Our route on this part lay along the watershed between Lakes Nyassa and Leopold.

At Mtinga the land makes once more a sudden rise, forming the Chingambo Mountains, 6000 feet in height. This range is especially interesting, as separating the drainage of three different lakes and two great continental river systems. To the east we find streams flowing to Lake Nyassa, and also to Lake Leopold. To the west the streams drain into the Chambeze, and thence into Lake Bangweolo. The river system of the Zambesi is represented by the streams which flow to Nyassa, and that of the Congo by those which flow to Bangweolo. As Lake Leopold has no outlet, the streams which enter it must be considered to be distinct from any of the African great river systems.

On the western side the Chingambo Mountains dip gently down till the altitude is only 4500 feet at Mswilo, the chief town of Inyamwanga. From this place to Sombe's the ground undulates gently, but gradually rises in altitude till Sombe's is reached, where the Plateau again attains a height of 6000 feet, though many of the scattered hills around will be quite 7000 feet high. Along this line we still find ourselves on a watershed dividing the streams which flow to Lake Leopold from those which go to Bangweolo. On reaching Sombe's we approach Tanganyika, into which all the other streams fall.

Such is the general contour of this tract of country; on the whole it is marked by utter monotony in form and colour. There is not a bit of scenery, except at the two extremities, worthy of a single line of description; not a picturesque hill or lovely valley to relieve the landscape. Such a thing as a patch of primeval forest does not exist, and nowhere is a large tree or profuse vegetation seen, except perhaps on specially favourable points by the bank of some stream where a little rich alluvium has been collected, and where moisture exists throughout the year. For days together the traveller passes through soul-wearying stretches of scraggy bushes and trees, none thicker in the stem than a man's leg, and most of them barked and cut. The only comfort to be derived from our recollection is the fact that there were at least no marshes and swamps, and no mosquitoes.

A few words will suffice to describe the geological character of this region. At Makula's we have already noticed the occurrence of a great series of volcanic rocks, probably belonging to the Tertiary period, but exhibiting signs of minor eruptions of a vastly later date, if not absolutely recent in a geological sense. From Makula's we find the rocks to be of the same nature as those whose acquaintance we made in Uhehe, Ubena, and Upangwa,—that is, metamorphic clay-slates, schists, gneiss, and even granite at a few points. Round the south end of Tanganyika a sudden change takes place in the rocks. On leaving Sombe's we pass in a few steps from these ancient formations to those of vastly later age—carboniferous red and grey sandstones and quartzites, much broken and jointed, but still retaining their original horizontality.

Regarding the natives little need be said. We have already described in some detail the people of Makula, and enough has been seen to indicate the existence of a radical difference between them and the tribes on the higher Plateau. Between these latter—the Wanyika, Wanyamwanga, Wamambwe, and Walungu—no strict line of separation exists. It is true if we compare the Walungu with the eastern Wanyika we shall find very marked differences in appearance and customs, but when traced from east to west through the intervening tribes an insensible gradation is found to exist, so that at no one point do we observe a separation between the people further than in government and name. As we have already seen, the eastern Wanyika are an exceedingly rude and warlike race, with much hair about their face and body, living independently in small filthy villages, and relying for sustenance almost entirely upon the few cattle they possess. The Walungu, on the other hand, are lighter coloured and better featured, with less hair on their face and body, living in villages of considerable size under chiefs of some power, having no cattle, and depending entirely upon agriculture for subsistence. They are also much more cowardly, and having been brought a good deal in contact with traders are possessed of more cloth and ornaments.

Throughout Nyika no grain is cultivated except the indigestible and heating ulizè. Of vegetables, sweet potatoes are grown. In Inyamwanga a little millet and Indian corn is produced in favourable situations, though the ulizè still remains the staple food. In Ulungu the supplies of food become more varied. Cassava is added to those previously mentioned. Tobacco is abundant, and cotton, though not very abundant, is found everywhere, and of it the poorer natives make a very coarse cloth.

It will be remembered that in describing the Mahenge we traced the cause of their sudden and extraordinary rise into a powerful tribe. The Walungu present a similar case of the great influence of a name and a dress in Africa.

It appears that about twenty years ago, perhaps more, Ulungu was governed by a chief called Kakungu. At that time his people were much harassed by the constant raids of Mazitu—or, as they were known here, the Watuta—under a chief called Tafuna. During one of these raids, Mulalami, Kakungu's eldest son, was captured and carried off as a prisoner by the Watuta. He remained with them several years, and thoroughly learned their customs and mode of warfare, with which he appears to have been impressed. At last he either escaped or was allowed to go free, and he returned to his own country, Ulungu. He at once commenced and drilled the Walungu into the Watuta mode of warfare, causing them to adopt the same dress, arms, war-cry, and manœuvres, and soon they appeared real Watuta in everything except in origin. But the name and appearance was everything with the cowardly natives. At Tanganyika, as beside the Rufiji, the sight of a Watuta dress or cry was sufficient to carry fear into the hearts of men and disarm resistance.

The counterfeit Watuta took full advantage of this, and devastated the whole surrounding country. Kakungu becoming old, divided his kingdom between his five sons. Kasonso and Kapambe occupied the south-western district, Kitimba the western, Sombe the southern, and Mulalami the eastern, between Sombe's and Fipa.

At last Mulalami, who was the greatest of them all, and who had converted a tribe of peaceful agriculturists into one of ferocious warriors, died. His whole military system fell into ruins, and the people reverted to their former habits, dress, and arms, and now appear as most harmless natives, discarding the very name of Watuta. In every hut, however, is to be seen the oxhide shield, now never used, but kept as a relic of former military times. The spear and the assegai, and the savage dress, have been replaced by the bow, which they prefer to carry, and the simple cotton loin cloth or goat's skin.

Except at one or two places where they have been corrupted by contact with the traders, they are now the most gentle and polite of people. Their ceremonious salutations in the morning are surprising. On every side a continuous clapping of hands goes on, with the accompaniment of

"Kwi-tata! Kwi-tata!" which is their mode of saying, "How d'ye do?" As each person appears outside of his house, he needs must turn to every one singly, and, bowing politely, clap hands with the accompanying words. If a relative or some very great friend appears, they at once rush into each other's arms; while if a chief passes, they drop on their knees, bow their head to the ground, clap vigorously, and humbly mutter, "Kwitata! Kwitata!"

The Czar of Russia, the German Emperor, or any other of the continental potentates whose lives are frequently in danger, might get a very good hint from the Walungu as to the best way to keep themselves safe from conspirators, and secure the utmost vigilance on the part of their chief minister.

It appears that the headman or factotum of a Walungu chief is always killed on the death of his master, and buried with him. This is carried out by the people themselves, and is strictly adhered to as a national custom. Hence no Walungu headman ever dares to conspire against the life of his chief, as that would seal his own doom; and it is equally for his interest to ward off any attempts made by others, and be zealous for his safe keeping.

A striking trait in the character of the Walungu is worth mentioning. I heard of it during my stay among the Arabs of Unyanyembe. Slaves brought from Ulungu become perfectly unmanageable, dull, and morose, and in a very large number of cases they commit suicide—an act rarely perpetrated by natives of any other tribe. And yet as slaves in Unyanyembe, they are vastly better off both in food and clothing than if they were free, besides having next to no work. It must therefore arise simply from grief at being separated from their country and kindred, and from a certain sense of the degradation attending slavery.

When a Walungu chief dies a most barbarous custom prevails. All his wives (and they are always numerous), with one exception, together with his headman, are killed, and buried with him. A worse fate awaits the one not killed. A hole, just sufficient to hold her, is dug. In this she is placed, and then covered over, leaving only a small aperture, through which she breathes, and by which a spear passes down into her hand. At the end of the second day of burial, if it is found that she has survived this horrible incarceration, and that she still holds the spear, she is taken out and allowed to live. If she be dead, or gives no sign, then there is no further need of burial, and so the matter is finished.

For Product Safety Concerns and Information please contact our EU representative GPSR@taylorandfrancis.com Taylor & Francis Verlag GmbH, Kaufingerstraße 24, 80331 München, Germany

Batch number: 10725867

Printed by Printforce, the Netherlands